Joint Conference of Modern Materials and Manufacturing 2019

AA005372

12th International DAAAM Baltic Conference and 27th International Baltic Conference (BALTMATTRIB 2019)

Tallinn, Estonia
23 – 26 April 2019

Editor:

Tauno Otto

ISBN: 978-1-7138-1171-8

Printed from e-media with permission by:

Curran Associates, Inc.
57 Morehouse Lane
Red Hook, NY 12571

Some format issues inherent in the e-media version may also appear in this print version.

DAAAM International Vienna
Bahnallee 10
A-2120 Wolkersdorf
Austria, EU

Phone: +436646424398

secretary@daaam.info

Additional copies of this publication are available from:

Curran Associates, Inc.
57 Morehouse Lane
Red Hook, NY 12571 USA
Phone: 845-758-0400
Fax: 845-758-2633
Email: curran@proceedings.com
Web: www.proceedings.com

TABLE OF CONTENTS

DEVELOPMENT OF CYBER-PHYSICAL PRODUCTION SYSTEMS BASED ON MODELLING TECHNOLOGIES 348
K. Mahmood, T. Karaulova, T. Otto, E. Shevtshenko

INNOVATIVE METHODS OF ENGINEERING EDUCATION POPULARIZATION AT SCHOOLS 356
E. Shevtshenko, T. Karaulova, M. Igavens, G. Strods, I. Tandzegolskiene, V. Tutlys, K. Mahmood

SYNCHRONIZING PHYSICAL FACTORY AND ITS DIGITAL TWIN THROUGH AN IIOT MIDDLEWARE: A CASE STUDY 364
V. Kuts, G. Modoni, T. Otto, M. Sacco, T. Tahemaa, Y. Bondarenko, R. Wang

OPTIMIZATION OF ENTERPRISE ANALYSIS MODEL FOR KPI SELECTION 371
S. Kaganski, M. Eerme, E. Tungel

A SIMULATION STUDY TO DETERMINE THE PARAMETERS OF MEDICINE INVENTORY POLICY 376
L. Al-Fandi, A. Obaid, B. Alfailakawi, H. Alsubaiei, S. Khudhair

SUITABILITY ANALYSIS OF USING INDUSTRIAL ROBOTS IN MANUFACTURING 383
T. Kangru, J. Riives, K. Mahmood, T. Otto

INTEGRATION OF AUTONOMOUS VEHICLES AND INDUSTRY 4.0 389
R. Sell, A. Rassolkin, R. Wang, T. Otto

MODULAR SMART CONTROL SYSTEM ARCHITECTURE FOR THE MOBILE ROBOT PLATFORM 395
R. Sell, E. Valjaots, T. Pataraia, E. Malayjerdi

PRODUCTION MONITORING SYSTEM DEVELOPMENT FOR MANUFACTURING PROCESSES OF PHOTOVOLTAIC MODULES 401
P. Tsukrejev, K. Kruuser, K. Karjust

IMPLEMENTATION OF A KNOWLEDGE-BASED MANUFACTURING ON THE EXAMPLE OF SUMAR TOOLS OU 407
K. Kruuser, J. Riives, P. Tsukrejev, I. Kiolein

METHODOLOGY FOR IMPLEMENTING UNIVERSAL GRIPPING SOLUTION FOR ROBOT APPLICATION 413
M. Azim, A. Lobov, A. Pastukhov

GRANULAR JAMMING BASED ROBOTIC GRIPPER FOR HEAVY OBJECTS 421
J. Miettinen, P. Frilund, I. Vuorinen, P. Kuosmanen, P. Kiviluoma

Author Index

Pagination in this book matches the original digital media

Proceedings of the Estonian Academy of Sciences,
2019, **68**, 4, 347
Available online at www.eap.ee/proceedings

EDITORIAL

Foreword

Over the past 23 years, DAAAM Baltic has been an international forum for researchers and engineers to present their research results in the areas of industrial engineering, manufacturing and automation. During these years its scope has been expanded from design engineering, production engineering and production management to mechatronics, IoT, digital twins and additive manufacturing. This year the 12th International DAAAM Baltic conference was organized jointly with the 27th International Baltic Conference BALTMATTRIB 2019 as joint conference of Modern Materials and Manufacturing 2019. It brought together researchers from the academic communities and industry to share their experiences, to foster collaboration, and to explore the trends in materials engineering, in product and manufacturing development, based on the use of IT and computer systems. The DAAAM Baltic focus encompasses theoretical and practical methodologies, resolving technologies of CAD, CAE, MPC and many others, considering completeness, dynamics of change, new technology insertion, complexity, integration, and time to market. Our society needs engineers and managers having technological knowledge, global vision and professional management knowledge in order to meet the challenges of the new age. We emphasize the importance of European initiatives towards *Factories of the Future* and technology platform *Manufuture*, and hope to give our contribution by bringing scientists, entrepreneurs and industry specialists together for exchange of future visions based on the scientific researches and case studies.

DAAAM Baltic meetings are regional events organized under auspices of Danube Adria Association for Automation and Manufacturing (DAAAM) in co-operation with DAAAM International Vienna, Tallinn University of Technology (TalTech), Aalto University, Estonian Academy of Sciences, BALTECH Consortium, IMECC, Federation of Estonian Engineering Industry, Association of Estonian Mechanical Engineers, Smart Industry Centre (SmartIC) and Tallinn City Enterprise Board. This special issue contains selected papers from the authors, presented at the conference Modern Materials and Manufacturing 2019, hosting the 12th International DAAAM Baltic Conference.

Tauno Otto
Guest Editor

Proceedings of the Estonian Academy of Sciences,
2019, **68**, 4, 348–355
https://doi.org/10.3176/proc.2019.4.02
Available online at www.eap.ee/proceedings

MANUFACTURING
ENGINEERING

Development of cyber-physical production systems based on modelling technologies

Kashif Mahmood*, Tatjana Karaulova, Tauno Otto, and Eduard Shevtshenko

Department of Mechanical and Industrial Engineering, Tallinn University of Technology, Ehitajate tee 5, 19086 Tallinn, Estonia

Received 12 April 2019, accepted 2 May 2019, available online 26 September 2019

Abstract. Internet of Things (IoT) is extending quite rapidly into the physical and virtual manufacturing worlds. IoT facilitates digitalization of manufacturing systems, which are considered vitally important to enhance the effectiveness and efficiency in the future manufacturing era. In order to compete globally and to meet the rapid market changes, manufacturing companies should consider implementation of manufacturing systems that are self-organized and decreasing constant human intervention to a minimum, still keeping the process under human control. In this paper, authors used the concept of Industry 4.0 to upgrade the manufacturing system according to the modern manufacturing needs. Manufacturing systems of Industry 4.0 are called cyber-physical production systems (CPPSs). On the other hand, modelling technologies such as process and data modelling, help to establish and understand the performance of a production system, as the evaluation of the control of an automated production system helps to make it reconfigurable and leads to grasping the idea behind the control mechanism of a CPPS. The aim of the paper is to develop a CPPS based on modelling technologies and to propose a concept of upgrading a relatively traditional production system.

Key words: CPPS, process and data modelling, manufacturing systems, IoT.

1. INTRODUCTION

Nowadays, the manufacturing industry is facing significant challenges. The ever-growing consumer market demands new, sophisticated and high-quality products while retaining their low price and short production times. Also, as compared to what the situation was twenty years ago, product lifecycles have decreased dramatically, which forces the manufacturing companies to discover new and more efficient ways of producing goods and offering services [1]. Production lines of the future would not be easily upgraded, they should rather upgrade themselves. Moreover, Industry 4.0 paradigm interconnects the parts of a production line, thanks to embedded systems, i.e. tiny computers, sensors and actuators built into the manufacturing system, forming an industrial IoT. The manufacturer is usually slow in implementing new technologies and accepting new approaches when doing ordinary tasks, which potentially results in substantial monetary losses. Already today, by implementing newer technology and techniques, production could be made faster and more efficient, resulting in savings for both – for companies and their customers, and in decreasing our impact on the environment. As humanity moves towards total reusability and environmental friendliness, our vehicles become electric, our rockets are designed to fly for dozens of times and we try to recycle materials as much as possible.

* Corresponding author, kashif.mahmood@taltech.ee

Companies do not build different factories to produce products with different shape and size. Instead, the production line should be updated to fit new requirements. Therefore, production lines of the future should not only be easily reconfigurable but they should also update themselves. It is, therefore, logical for the manufacturing itself to become flexible and rapidly reconfigurable.

Cyber-physical systems (CPSs) are the systems of collaborating computational entities, which are interconnected with the surrounding physical world. Those on-going processes simultaneously use data acquisition and data-processing services and make it available on the Internet [2,3]. Authors suggest solving the existing problems using CPS and implementation of IoT in the manufacturing systems. Authors use a case study to develop the concept and to show its relevance.

In this study, the production system was analysed. The production system was divided into modules and analysis were performed based on the functionalities of the system's components. Therefore, the authors used modelling techniques based on Integrated Definition (IDEF) standard. Futhermore, the study describes a generic approach how to digitalize an outdated (traditional) manufacturing system. A production system was selected to grasp the relevance of the generic approach and to make the studied system rapid according to the modern manufacturing needs, i.e. a reconfigurable manufacturing system (RMS) and ultimately a cyber-physical production system (CPPS). The structure of the study can be depicted in Fig. 1, which shows the tasks and methods used for the manufacturing system.

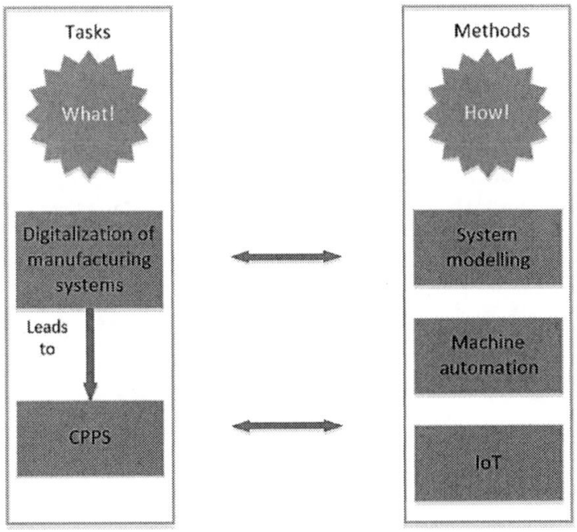

Fig. 1. Structure of the study.

2. LITERATURE REVIEW

In this section, the authors give a brief overview of the tasks and methods, such as modelling technique system, the role of automation in advancement of manufacturing system, the importance of distributed manufacturing systems, the concept of IoT and the background of CPPS. Several articles and manuscripts of the related fields are used to compile the review of the literature.

2.1. System modelling

Modelling of a system defines the process of establishing steps or workflow of a particular system that is selected for the study. Process modelling plays a significant role in identifying, establishing, analysing and extracting the activities of a system. It fulfills the requirements of transforming business process, manufacturing process, and product-development process into a process view model [4]. Massive amount of data and system complexity can lead to confusion. Therefore, a system study always requires activity modelling. It is usually challenging to recognize and remember all information, which should be known about the systems [5]. Selection of right key performance indicators (KPIs) has also an impact on a production system analysis and monitoring [6]. The authors used modelling techniques based on Integrated Definition (IDEF) standard. IDEF refers to a family of a modelling language, which covers a wide range of uses, from functional modelling to data simulation [7].

2.2. Machine automation and manufacturing systems

Automation is the use of mostly automatic equipment in a system of manufacturing or another production process [8]. There is one thing that has changed the concept of an automated machine and defined the understanding of what automation is in the modern world. That was the invention of the computer. For the first time in history, humanity has invented such a sophisticated and powerful tool, capable of performing unbelievable amounts of operations over the short periods. Over the last few decades, this technology has advanced even further. Already today, there are computers able to think, create and make decisions [9]. Moreover, as it seems to be, they are much better and faster than humans are, which brings a new understanding of automation.

A manufacturing system defines the organization to produce a product. There are many different types of manufacturing systems that came into being in the history of manufacturing due to the advancement of technologies and market fluctuation [10]. The development of manufacturing systems started from the concept of assembly line, followed by Toyota Production System (TPS), then flexible manufacturing system (FMS) and reconfigurable manufacturing system (RMS), after that web- and agent-based manufacturing system, nowadays cloud-based manufacturing system and to the recent future CPPS [11].

2.3. Concept of IoT

The exposure of the Internet has altered the daily business and personal lives during past years. It is likely to continue also in the future as it can be depicted from the current trends. IoT is an extension of the Internet. It provides instant access to the information about physical objects and leads to innovative services with high efficiency and productivity [12]. IoT adoption leads to changing old manufacturing systems into modern digitalized ones, where they are enabled to interact and communicate between themselves and with the environment by exchanging data and information "sensed" about the environment while reacting autonomously to the "real/physical world" events and influencing it by running processes that trigger actions and create services with or without direct human intervention [13]. The National Institute of Standard and Technology (NIST) describes "Cyber-physical system as the IoT, which involves connecting smart devices and systems in diverse sectors like transportation, energy, manufacturing and healthcare in fundamentally new ways. Smart cities/ communities are increasingly adopting CPS/IoT technologies to enhance the efficiency and sustainability of their operation and improve the quality of life" [14].

The Multi-Scale Systems Engineering Research Group at Georgia Institute of Technology is developing various IoT and cyber-manufacturing related technologies. For instance, big data analytics has been applied to predict manufacturing costs in cyber-manufacturing [15].

2.4. Overview of CPPS

The manufacture-orientated CPS is also called CPPS, the bases of CPS are highlighted by [16] as the representation of a physical system in the virtual world and the virtual system can be presented in the physical world. The cyber-physical systems are an expansion of embedded systems, where sensors are used to record, save and evaluate data. They interact with the physical and virtual world, communicate each other through interfaces, and use big data and human-machine interfaces [17]. CPS support real processes and provide operational control of objects – the best way to do it on a base of the process model. Process model directly manages the data and business logic of the application. The controller handles communication between users and a model. This layer accepts inputs and converts it into commands for the execution of business logic.

The components of CPPS should have their information and consist of structure information storage. The information can be exchanged between the components via communication interfaces of the CPPS and can be adopted for autonomous decisions in the production process. The communication or process control in a CPPS can be executed directly among the components as shown in Fig. 2, where the control system does not include a central or supervisory control at all. Instead, all the direct process controllers are supposed to interact with each other using standard industrial Ethernet protocols. From the software point of view, these direct process controllers should interact with each other as remote input/outputs (I/O) with no particular set up required to have everything in synchronization [18].

3. METHODOLOGY AND STEPS FOR CPPS IMPLEMENTATION

The primary method used in this paper is the case study approach to define and realize the digitalization of a manufacturing system based on the IoT concept and modelling techniques. Moreover, the literature review was carried out to describe and understand the fundamental techniques that help to build an approach of digitalization. This study defines an approach to examine and digitalize a manufacturing system based on

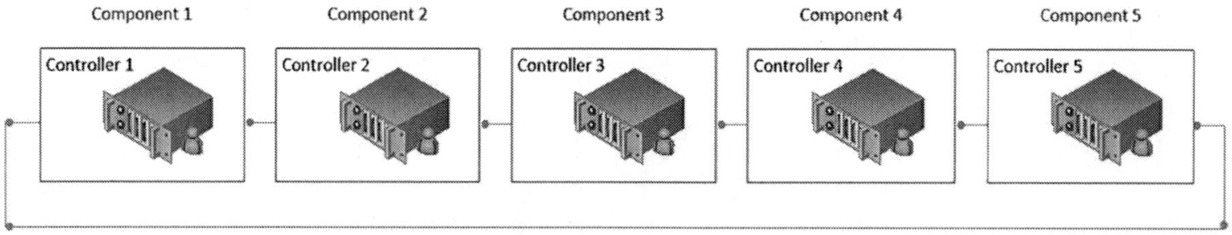

Fig. 2. Process control in a CPPS [18].

modelling technologies and CPS applications. Proposed steps are as follows:

1. Determining rationale and obstacles;
2. Designing the control system architecture;
3. Process modelling and data modelling of the system;
4. Implementation of CPS enablers;
5. Establishing communication network between all devices and components;
6. Installation of smart sensors for data collection;
7. Data processing through intelligent data analysis algorithm;
8. Visualization the status via a web browser (dashboard).

The conceptual model as shown in Fig. 3 has introduced the general approach to redesigning any existing automatic manufacturing system to CPPS. The core of this work is a process model, which directly manages the data and process logic, business rules and application data. A central element of every process model is activity or operation. Operation analysis is well-studied and widely used in industry and is connected with several types of reliability analysis, e.g., failure modes and effects analysis (FMEA), or any other, which have the similarity with process model structure, e.g., reliability block diagram (RBD) or fault tree analysis (FTA). Reliability data also may be included in the process model and used in the logical model of system controllers. In the current research, a database (DB) is used for the process time regulation, which collects process flow information and is the basis for process correction. Structure of DB is also formed by the process model. Since the process model contains the time parameter for every process activity, it is reasonable to use process simulation analysis for controlling the process changes, which significantly reduce the possibility of process errors.

4. CASE STUDY

4.1. Description of the system and obstacles

The manufacturing system was considered as a test-case to verify the proposed approach. The system is ususally used for academic research and practical purposes. The system was outdated, slow and difficult to control with the modern tools. Besides, most of the components were not working properly. The system consists of the individual components, and the layout of the system is depicted in Fig. 4.

The proposed changes in the manufacturing system were mainly related to the controllers, each component would have microcontroller. Other than those, the analogue height and width measurement gauges were replaced with time-of-flight laser sensors. It was suggested that both robotic arms of the system should be of the same model and characteristics, as well as the conveyors. The controllers' description and communication are described in the later section.

4.2. Control system architecture

Existing control system of the test manufacturing system resembled the traditional central control system.

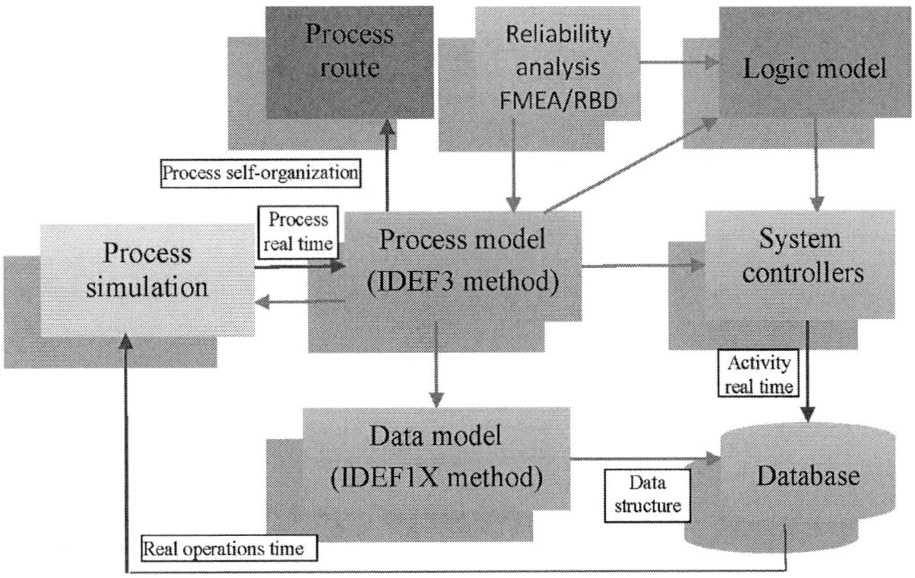

Fig. 3. Modelling centred approach for process self-organization.

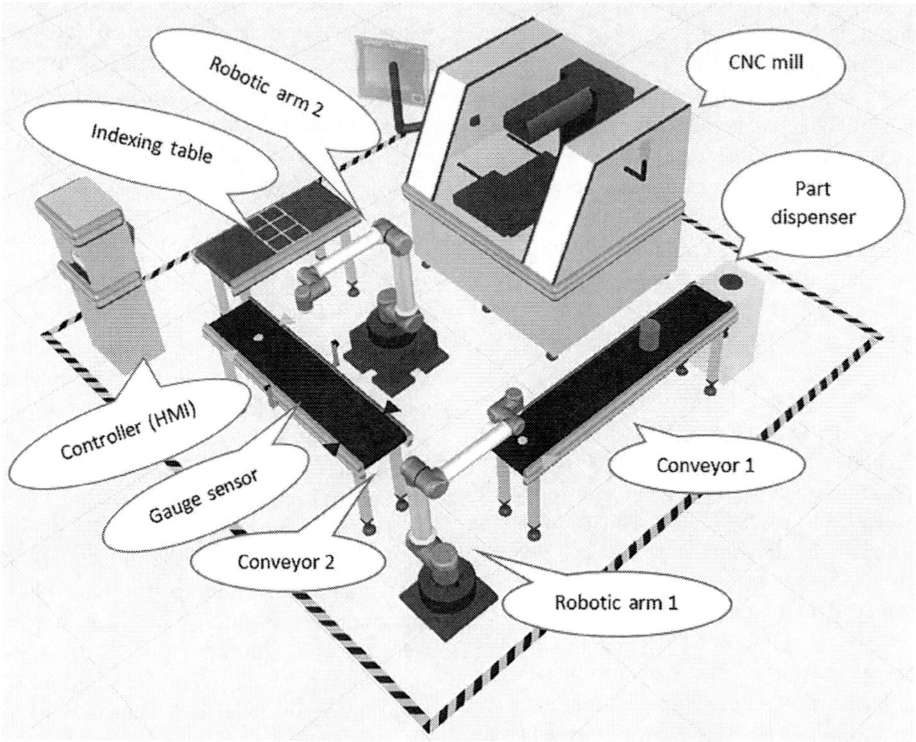

Fig. 4. Manufacturing system components and layout.

The central controller is directly or indirectly connected to the internal controller of every individual component of the system. The information flow and hierarchy can be depicted in Fig. 5.

Proposed control architecture for the test system is the combination of traditional hierarchical control with the distributed control as defined in the Industry 4.0 paradigm. The direct process controller of an individual component of the system communicates straight with controller of other components. Furthermore, the process controller of each component should also communicate with supervisory control of the system. The purpose of the supervisory control in the suggested architecture is to act as human-machine interface (HMI), data acquisition unit and data server. Therefore, it leads the system into the IoT domain and CPPS category. The distributed control and transmission of data from component to component results in faster processing speed of the system. The separate role of each component of the manufacturing test system allows reprogramming components individually. The new control system architecture can be seen in Fig. 6. Arduino microcontrollers (Nano, Uno and Mega) were used with Raspberry Pi3 as a single board computer (SBC) to collect data from microcontrollers and act as an HMI for CNC mill.

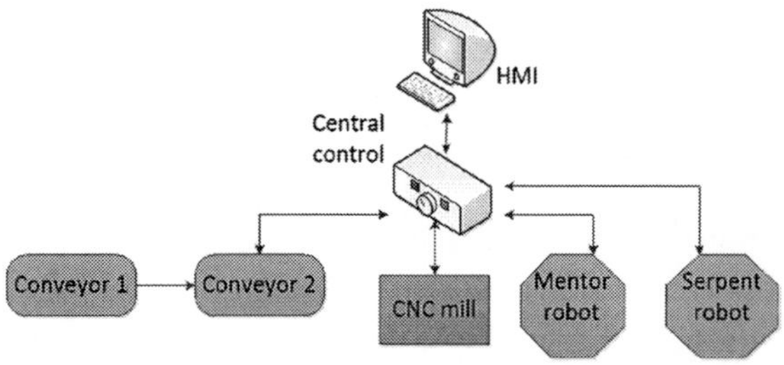

Fig. 5. The general control hierarchy of the manufacturing system (As-Is).

4.3. Modelling of the manufacturing system

The process modelling was performed by the IDEF3 method. The process model showed in Fig. 7 contains the same reference data and activity flow: arrows with double heads show the work-part movements. Arrows with single head show precedence and connections, dotted arrows show information flowing to the database (DB).

Database structure was elaborated by using the IDEF1x method as shown in Fig. 8. Every model object has a table of relation with DB and may contain many

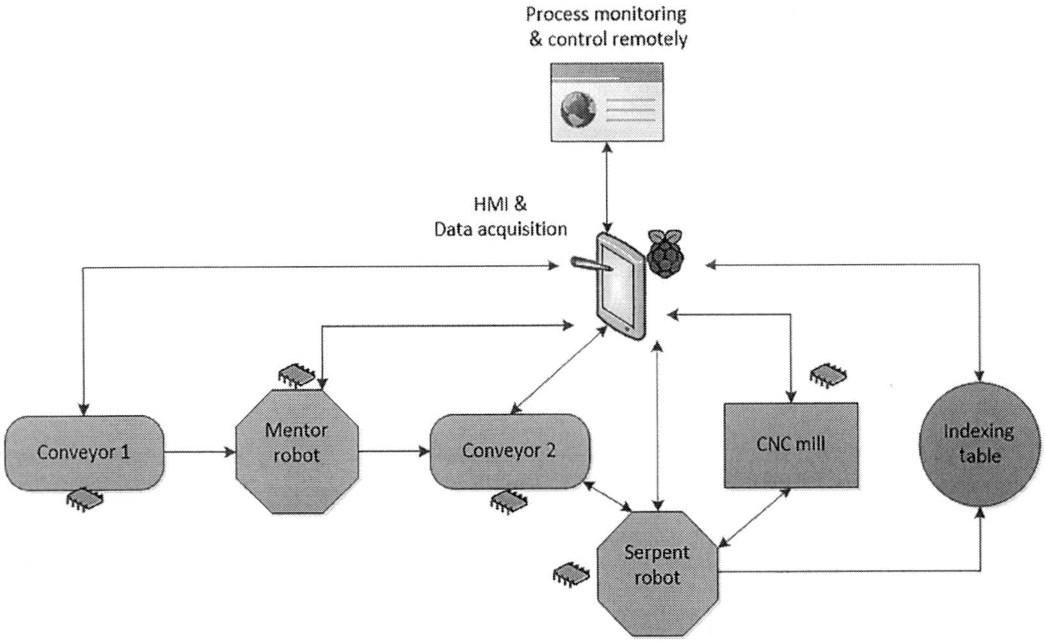

Fig. 6. The proposed control system architecture for the manufacturing test system.

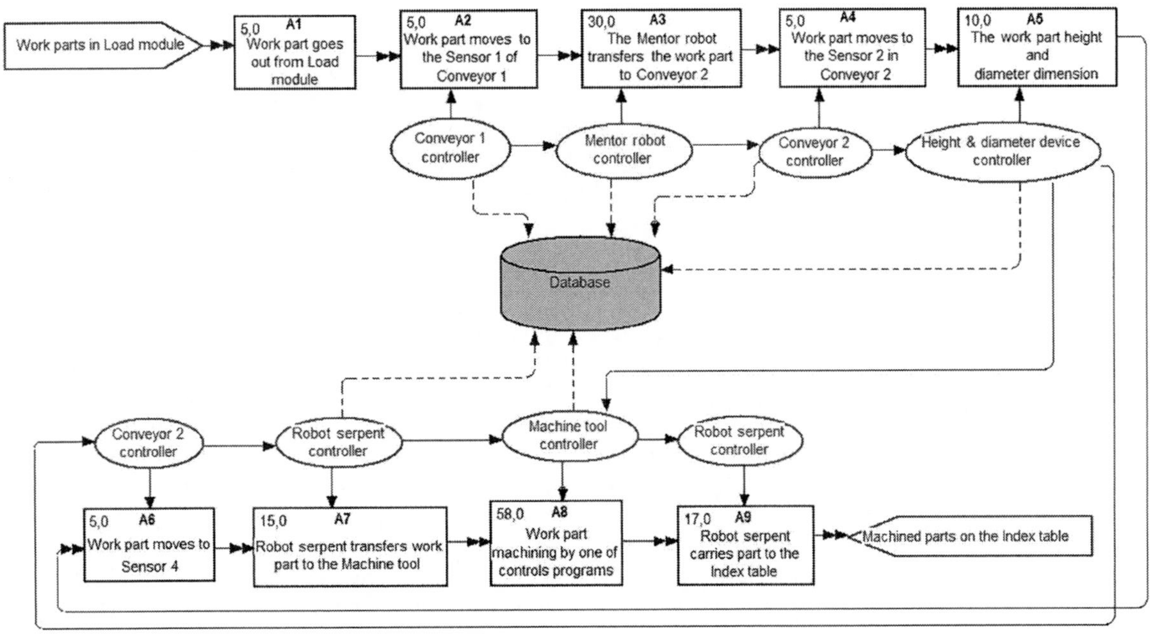

Fig. 7. The process model of the manufacturing system.

records. Since the database is orientated in process time regulation, the system objects contain only attributes which affect the function duration. The yellow marked process activities (A1, A2–A9) reflected the start time and end time tags of each activity. This IDEF3 model includes the duration of every activity, simulation of the model contains the activities duration deviations according to certain laws of distribution. Precise finish-

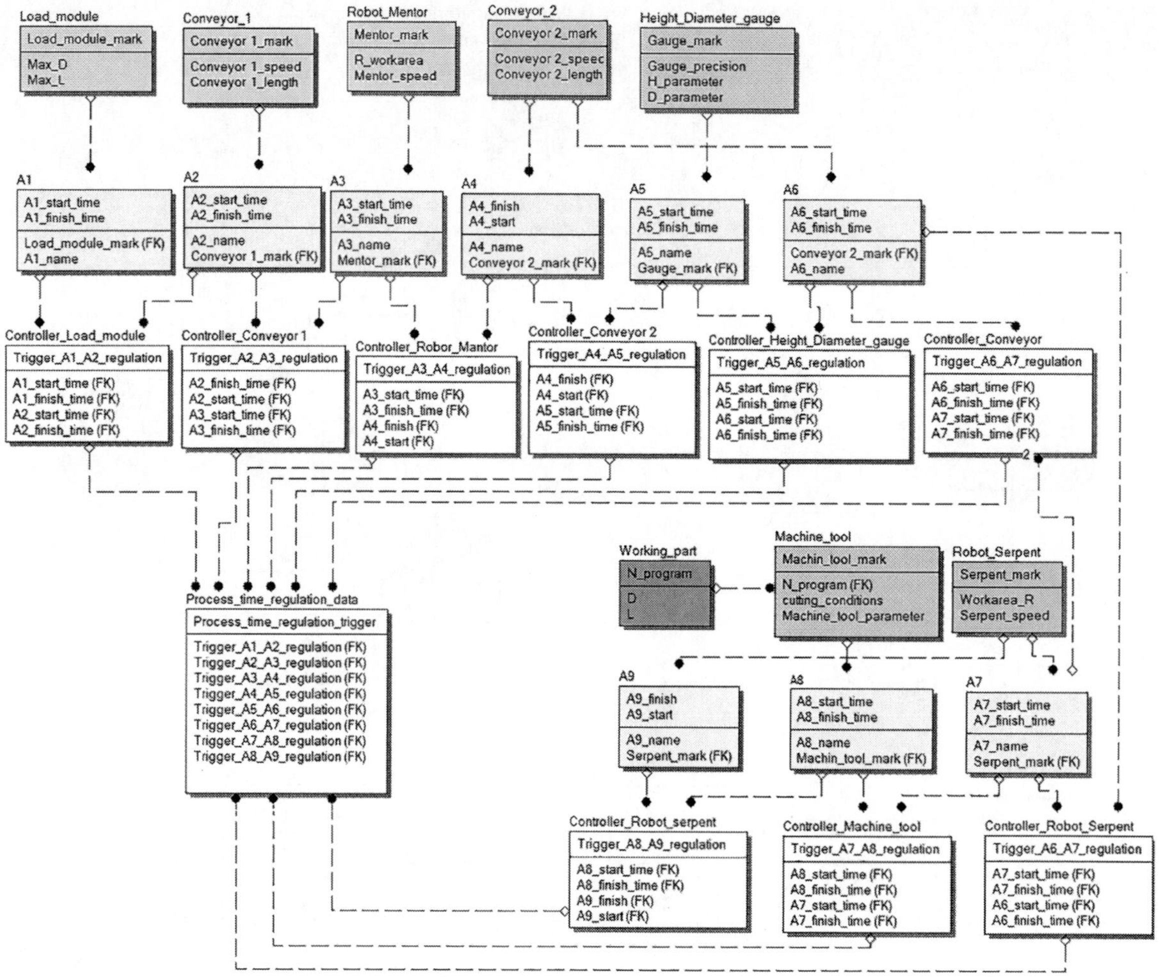

Fig. 8. The database structure of the system.

time and start-time of the next activity can be regulated only by controllers. It means:

$$\text{start-time\_A2} - \text{finish-time\_A1} \rightarrow 0.$$

The tables of the object controllers include the time attributes from previous and following activities. Process time regulation triggers calculate the time intervals for common activity time. The parameters showed in the DB are connected with time regulation.

5. DISCUSSION AND CONCLUSION

The overall improvement in the operating time of the system was observed as the inclusion of smart gauge sensor cut down the activity time used for measuring dimensions of the parts. Thus, the whole process would be balanced if the machining activity time could be decreased. This could be achieved through the appropriate cutting tool such as a solid end mill of steel alloy. Furthermore, the other main goal is self-organizing of the process route, if one component of the system malfunctioned the other components should figure out, how to complete the work cycle, e.g., if the mentor robotic arm stopped working then the serpent, robotic arm would do the additional activity of picking and placing to complete the cycle. Implementation of distributed control system enabled to communicate with some manufacturing execution systems (MESs), hence the system activities can be monitored at the higher. The components such as conveyors, can be reprogrammed to do anything they are physically capable of and they can perform the job with superior control, faster speed and with better adaptability.

All in all, in this paper digitalization and upgrading an old manufacturing system was developed, which is based on the Industrial Internet of Things (IIoT). A model of redesigning existing automatic manufacturing system to CPPS through process and data modelling was described. The proposed method was partially implemented to a manufacturing test system that was used for academic purpose case studies. Moreover, the proposed approach enables better planning, controlling and monitoring of the system, and that leads to proactive decision making. The detail implementation process can be extended and described in future work with a further demonstration of the system reliability analysis.

ACKNOWLEDGEMENTS

Special thanks and gratitude to the students of Integrated Engineering program, Department of Mechanical and Industrial Engineering, Tallinn University of Technology. This research was supported by project AR16077 Smart Industry Centre (SmartIC) No. 2014-2020.4.01.16-0183 and also supported by the EU Regional Development Fund. The publication costs of this article were covered by the Estonian Academy of Sciences.

REFERENCES

1. Paavel, M., Karjust, K., and Majak, J. Development of a product lifecycle management model based on the fuzzy analytic hierarchy process. *Proc. Est. Acad. Sci.*, 2017, **66**(3), 279–286.
2. Monostori, L. Cyber-physical production systems: roots, expectations and R&D challenges. *Procedia CIRP*, 2014, **17**, 9–13.
3. Lee, J., Bagheri, B., and Kao, H. A. A Cyber-Physical Systems architecture for Industry 4.0-based manufacturing systems. *Manuf. Lett.*, 2015, **3**, 18–23.
4. Cellier, F. E. and Greifender, J. *Continuous System Modeling*. Int. Edition, Springer Science+ Business Media, New York,1991.
5. Soshnikov, D. and Dubovik, S. Knowledge-based business process modeling and simulation. In *Proceedings of the 6th International Workshop on Computer Science and Information Technologies CSIT, Budapest, Hungary, 2004*, 169–176.
6. Kaganski, S., Majak, J., Karjust, K., and Toompalu, S. Implementation of key performance indicators selection model as part of the Enterprise Analysis Model. *Procedia CIRP*, 2017, **63**, 283–288.
7. Mahmood, K., Karaulova, T., Otto, T., and Shevtshenko, E. Performance Analysis of a Flexible Manufacturing System (FMS). *Procedia CIRP*, 2017, **63**, 424–429.
8. Groover, M. P. *Automation, Production Systems, and Computer-Integrated Manufacturing*, 3rd ed. Pearson, London, 2015.
9. Noble, D. F. *Forces of Production: A Social History of Industrial Automation*. Transaction Publishers, New York, 2011.
10. Tolio, T., Ceglarek, D., ElMaraghy, H. A., Fischer, A., Hu, S. J., Laperričre, L. et al., SPECIES-Co-evolution of products, processes and production systems. *CIRP Ann.-Manuf. Technol.*, 2010, **59**(2), 672–693.
11. Wu, D., Rosen, D. W., Wang, L., and Schaefer, D. Cloud-based design and manufacturing: A new paradigm in digital manufacturing and design innovation. *Comput.-Aided Des.*, 2015, **59**, 1–14.
12. Bi, Z., Xu, L. D., and Wang, C. Internet of Things for enterprise systems of modern manufacturing. *IEEE Trans. Ind. Inf.*, 2014, **10**, 1537–1546.
13. Mourtzis, D., Vlachou, E., and Milas, N. Industrial Big Data as a result of IoT adoption in manufacturing. *Procedia CIRP*, 2015, **55**, 290–295.
14. National Institute of Standards and Technology (NIST), 2014. https://www.nist.gov/el/cyber-physical-systems (accessed 2017-03-20).
15. Chan, S., Lu, Y, and Wang, Y. Data-driven cost estimation for additive manufacturing in cyber manufacturing. *J. Manuf. Syst.*, 2018, **46**, 115–126.
16. Lee, E. A. and Seshia, S. A. *Introduction to Embedded Systems. A Cyber-Physical Systems Approach, 2nd Edition*. MIT Press, London, 2017.
17. Kuts, V., Otto, T., Tähemaa, T., Bukhari, K., and Pataraia, T. Adaptive industrial robots using machine vision. In *Proceedings of the ASME 2018 International Mechanical Engineering Congress and Exposition, Pittsburgh, November 9–15, 2018*, 1–8.
18. Pilz Australia Industrial Automation LP, The future of control systems and Industry 4.0. http: //www.processonline.com.au/content/industrial-networks-buses/article/the-future-of-control-systems-and-industry-4-0-1365495420 (accessed 2016-03-01).

Küberfüüsiliste tootmissüsteemide arendamine modelleerimistehnoloogiate põhjal

Kashif Mahmood, Tatjana Karaulova, Tauno Otto ja Eduard Shevtshenko

Asjade Internet ehk nutistu (IoT) tungib üsna kiiresti nii füüsilistesse kui ka virtuaalsetesse tootmiskeskkondadesse. IoT hõlbustab tootmissüsteemide digitaliseerimist ja seda peetakse tootmise tõhususe suurendamisel oluliseks teguriks. Et ülemaailmselt konkureerida ja kiirete turumuudatustega toime tulla, peaksid tootmisettevõtted orienteeruma tootmissüsteemidele, mis on iseorganiseeruvad ning vähendavad inimeste sekkumist miinimumini, hoides protsessi siiski inimkontrolli all. Käesolevas artiklis on kasutatud Tööstus 4.0 kontseptsiooni, mille mõistes nimetatakse tootmissüsteeme küberfüüsilisteks tootmissüsteemideks (CPPS).

Teisalt aitavad modelleerimistehnoloogiad, nagu protsesside ja andmete modelleerimine, tootmissüsteeme luua ja mõista nende toimimist, kuna automatiseeritud tootmissüsteemide juhtimise hindamine võimaldab need ümberkonfigureeritavaks muuta ning ühtlasi tõsta need CPPS-i kategooriasse. Uurimuse eesmärk on arendada modelleerimistehnoloogiatel põhinevat CPPS-i ja välja töötada kontseptsioon traditsioonilise tootmissüsteemi uuendamiseks.

Proceedings of the Estonian Academy of Sciences,
2019, **68**, 4, 356–363
https://doi.org/10.3176/proc.2019.4.01
Available online at www.eap.ee/proceedings

MANUFACTURING
ENGINEERING

Innovative methods of engineering education popularization at schools

Eduard Shevtshenko[a*], Tatjana Karaulova[a], Maris Igavens[b], Gunars Strods[b], Ilona Tandzegolskiene[c], Vidmantas Tutlys[c], and Kashif Mahmood[a]

[a] Department of Mechanical and Industrial Engineering, Tallinn University of Technology, Ehitajate tee 5, 19086 Tallinn, Estonia
[b] Rezekne Akademy of Technologies, Atbrivosanas aleja 90, 4600 Rezekne, Latvia
[c] Vytautas Magnus University, K. Donelaicio Street 58, LT-44248 Kaunas, Lithuania

Received 12 April 2019, accepted 14 May 2019, available online 26 September 2019

Abstract. In recent years, there is constant deficit of students in technical specialities in higher educational institutions, especially in industrial engineering area. To solve this issue, three universities of the Baltic region initiated the joint project “Innovative methods for implementing interdisciplinarity in career counselling (IMATEII)”. The aim of the project is to develop a new approach to prepare schoolchildren choosing appropriate profession through collective efforts of the university, schools and enterprises. Authors introduce the new approach based on the metal and machinery industry. Readers can apply the results of the current analysis in other geographical regions and different industrial sectors.

Key words: IMATEII, career education, design, technologies and economics competencies, machinery industry.

1. INTRODUCTION

After joining EU, students of the Baltic states can improve their professional level in other European countries. Often people leave their own country because they do not have complete information about the opportunities to obtain the appropriate speciality in their own country. Therefore, students usually break studies and leave their country seeking for suitable position abroad, if they do not have enough information related to all aspects of the selected profession. Schoolchildren have difficulties in orienting themselves in the professional area and in finding the way how they can realize themselves in interests of their own country. It is necessary to help them to solve one of the most difficult and important questions – “What do you want to become?” Appropriate choice of high school, regarding the future profession makes the education process complete and efficient. A significant role in professional orientation belongs to the higher education institutions.

Another major challenge is the need for further integration of cross-sectoral skills into the teaching and learning process [1]. An important task is to encourage young people to choose a profession in the fields of mathematics, science or technology because high demand for graduates in those fields is extremely important for further innovation and growth of the industries.

In the context of Latvia, Lithuania and Estonia, the fields of Wood-Forestry, Metal-Machinery, and Agriculture-Food are considered to be fast-growing industries. Number of trained specialists is insufficient, and the motivation and interest of potential employees should be increased by school career counsellors.

Adequate training of qualified specialists is possible only in close cooperation between schools, universities and enterprises. Educational programs of the universities

* Corresponding author, eduard.shevtsenko@taltech.ee

should correspond to the needs of the industrial companies developed in a particular region. Therefore, there is a need for constant monitoring of the competencies of the participants in education and production systems.

2. THE PROJECT "IMPLEMENTING INTERDISCIPLINARITY IN CAREER COUNSELLING"

The target of the project is to develop new (interdisciplinary) approach in career consulting, using the results of the project materials.

Audience would be teachers who are working with career issues, such as career consultants, class teachers, youth workers. Goal of the project is developing teachers' professional competence and popularization of engineering education in schools.

2.1. Career education

Career education is the planned provision of events, courses and programs at educational institutions that are helping learners to develop the skills that correspond to their interests, abilities and opportunities, to set their career goals and to manage their achievements. It provides knowledge and understanding of the world of work, its link with education, career planning and further education opportunities, and ensures active participation in one's working life [2].

A modern, competitive economy requires employees who have the skills, knowledge and attitudes that can be used in any work situation and who have the ability and desire to adapt and succeed in a changing world continuously. Therefore, the skills suitable for employment are an essential part of career management skills. Employment skills are defined as "the set of qualities, skills and knowledge that is needed by every participant in the labour market to ensure their effectiveness in the workplace – for their growth, for their employer and the economy of the country as a whole" [3].

When it comes to the professions of technical disciplines, it is essential to recognize the set of skills needed – the specific, technical and general work skills that are partly made certain and learned in general education, such as subjects of Design, Technology and Economics.

2.2. Project implementation

Researchers of three Baltic countries have divided several industrial domains like forestry and wood, metal and machinery, agriculture and food, between Latvian Rezekne Akademy of Technologies (RTA), Lithuanian Vytautas Magnus University, and Estonian Tallinn University of Technology (TalTech) for better project implementation.

The researchers set out the following objectives for the current project:

- to develop and implement innovative practices in career education at secondary schools;
- to provide secondary school teachers with career guidance competence based on interdisciplinarity among Design, Technology and Economics;
- to facilitate learners to make a motivated and targeted career choice in selected fields of industries.

2.2.1. Analysis of enterprises' needs

The first task of the project was to study the most demanded professional competencies in the Wood-Forestry, Metal-Machinery, and Agriculture-Food industries and elaborate recommendations for methodology of the professional development training for career counsellors on career guidance in secondary school. A questionnaire was used to collect information about employers' needs for their staff competencies in Design, Technology and Economics. The results of this work were introduced in previous research papers of the same authors [4–6].

Results show that competencies related to Design are more often repeated in all sectors of the industries. It leads to the conclusion that competencies related to the Design are more general. Competencies related to the Technology were also repeated in all fields of industries, but not as often as Design. Competencies related to the Economics repeated very rarely. That leads to the conclusion that competences related to the Economics are perceived to be more specific for every industrial sector. Probably the size of the companies in different industries varies, and that leads to the specific demands of competencies.

In their recommendations authors try to determine the general principles of how to connect guidance and secondary school students' competencies development related to specific industries. The results give opportunity to connect demand for competencies in Design, Technology and Economics with Teaching and Learning Strategies for the Critical Thinking three phases – Evocation/anticipation, Comprehension/the building knowledge, Reflection/consolidation [6].

2.2.2. Dissemination of engineering education in schools

The primary objective of this project is to attract and train students in schools to apply for technical vocations in higher educational institutions.

Figure 1 shows the primary collaboration between the project group members from the universities and schools.

Within the framework of this project, the primary emphasis is on familiarizing the students with the realities of the chosen profession through visits to the enterprises and master classes in school and university laboratories. It is essential to show the students how companies make the selected products in real production process. The specifics of technological and work processes of different industrial sectors that limit the observation of work processes should be taken into consideration, e.g., safety issues in heavy industries like metal and machinery, also noise, smell and heat, which are related to certain technological processes of production of different metal products. In some cases, it is hard to observe these production processes directly. In such case the observation of simulated or smaller scale technological processes can successfully replace the observation of real and big scale production. For example, students can visit the metal lab, which accepts the visitors and even offers possibility to have a "hands-on" practical experience in some production processes. It could improve understanding technological processes and manufacturing of metal products, compared to just visiting a large company and observing technological processes from a distance.

University lecturers and researchers prepare slides and training materials to give over to schoolteachers to improve their technological knowledge and skills. Schoolmasters, supported by the university lecturers, explain the schoolchildren how to design and manufacture the selected products, based on practical lessons. Also, project leaders must organize visits to the companies, which produce previously studied products, so that the schoolchildren have a complete overview of a real manufacturing process and production environment.

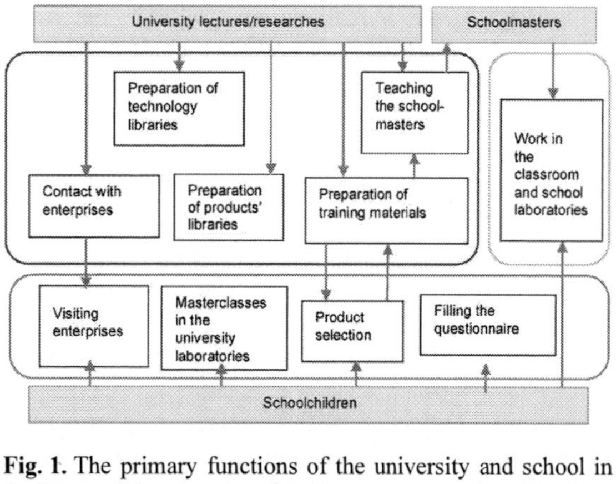

Fig. 1. The primary functions of the university and school in the frame of the current project.

It is essential because often working conditions and sensory experiences play a critical role in choosing a future profession.

Work in the classroom with schoolchildren must be based on teaching and learning strategies [7]. Authors give recommendations how to implement critical thinking, teaching and learning strategies in career guidance and how to introduce local industries to the pupils of secondary school through development of teachers competencies in Design, Technology and Economics.

The goal of the project is to implement the system of labour education and vocational guidance. The project develops the needs of young people in socially significant work activities, enables professional self-determination and social adaptation of schoolchildren adequate to their abilities.

2.2.3. The organizational activities of the project

This work results in elaborating teaching materials and preparation (training) of schoolmasters based on new methods and techniques developed during current project. Description on activities is introduced in Fig. 2.

The guidebook contains examples of how to implement the idea of the project, using critical thinking and learning strategies in career guidance, and how to allow secondary school students to get acquainted with local industries.

All materials were initially prepared in English, discussed in project team seminars and finally translated into Estonian, Latvian and Lithuanian. Schoolchildren do not easily understand new processes and new terminology. Therefore, teaching materials should be translated into all three languages of the Baltic states. Organizing the teachers' training of the three Baltic states, speaking different languages is not an easy task.

The next step was training the teachers to prepare them to carry out lessons in engineering classes in secondary schools. Slides and videos were prepared by the university lecturers to explain technological processes. Discussions how different products are produced were part of the practical work. Examples of slides made for the lessons are shown in Fig. 3.

2.3. The concept of the project

The concept of the project for each field of industry (Wood processing, Food, Mechanical engineering) is based on three phases of critical thinking: Design, Technology, and Economy (Fig. 4.).

The main idea in a constructivist approach is the student's thinking activity and the development of critical thinking. That requires a radical renewal of teaching

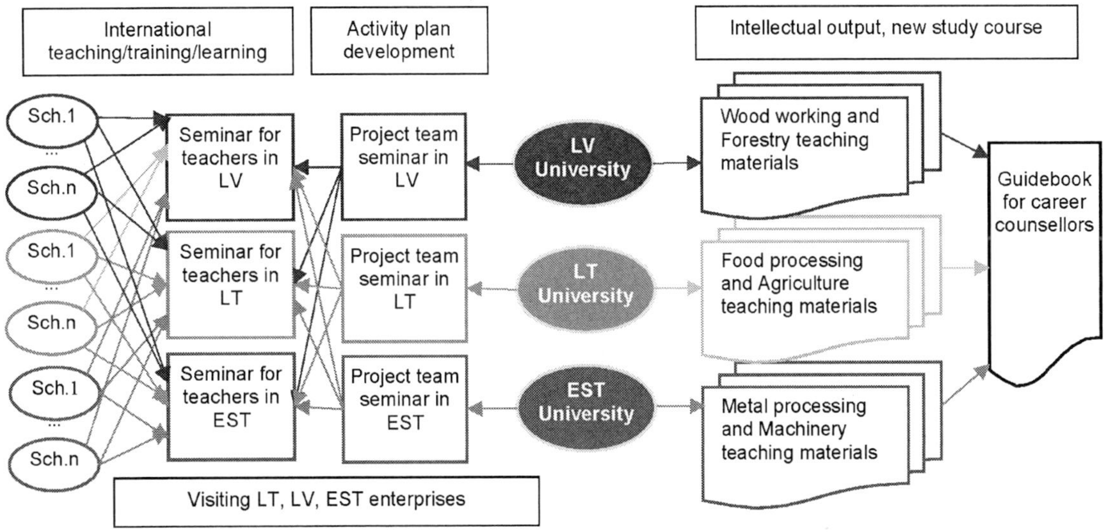

Fig. 2. Organizational activities for the project implementation.

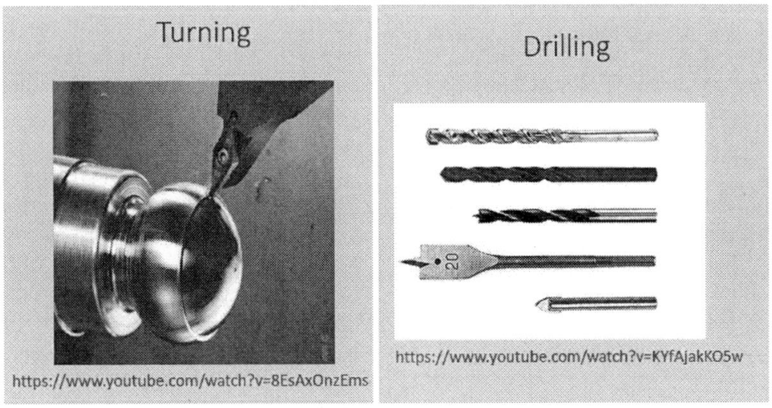

Fig. 3. Example of teaching materials.

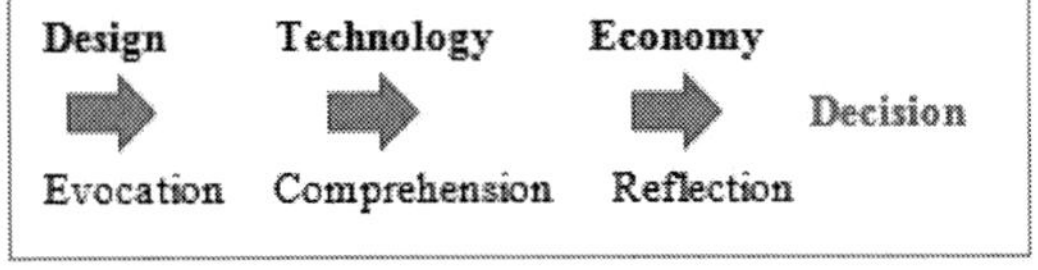

Fig. 4. The concept of the project.

methods and resources, and the teacher's ability to use diverse teaching techniques. It is essential for students to create an understanding of the succession and continuity of the learning process [8].

Learning process is structured in successive phases:

- evocation – the phase where educator uses structured pattern of questions to find out student's knowledge and skills and set learning goals;
- comprehension – the phase in which students are looking for a new knowledge and as a result of their actions, create their awareness and significance;
- reflection – the phase in which students look at the ideas they have learned and understand their meaning, ask questions, interpret, apply, discuss, test, and extend meaning by transferring it to other areas of activity.

For successful implementation of the system of principles, critical thinking offers innovative methodological solutions [9].

Critical thinking and problem solving skills include the ability of individuals to a) reason effectively, b) ask pointed questions and solve problems, c) analyse and evaluate alternative points of view, and d) reflect critically on decisions and processes [1].

Critical thinking is defined as the ability to analyse, interpret, evaluate, summarize, and synthesize information [10].

Each institution establishes a single module, which will be later integrated with other two parts of the curriculum.

The competencies of the Economy are: defining the most cost-effective solutions, analysing production processes, connections between different functions and departments, ability to see the whole picture of value chain from sales till support processes.

Based on this concept, Technological cards were developed for every considered type of industry [11–15]. Example of Technological card for metal cutting process is shown in Fig. 5.

2.4. Practical work for schoolchildren

Classes in schools are conducted by the schoolteachers with the support of university researchers and lecturers. Example of the practical work with schoolchildren is introduced in Fig. 6.

This example provides an opportunity to connect the competencies in Design, Technology and Economics with teaching and learning strategies using three phases of critical thinking – evocation/anticipation, comprehension/knowledge building, and reflection/consolidation [6].

2.4.1. Evocation/Anticipation phase – Design

At the beginning of the group counselling session the counsellor gathers the answers from schoolchildren to the following questions: What do you know about the machinery industry sectors? How much have you heard about it from the media (newspapers, magazines, TV, radio, the Internet (social networks, YouTube, company websites, etc)), family members, friends, neighbours? Have you done summer jobs in this field of industry? Have you volunteered in this industry? Have you visited a company in this industry? Do you have a plan after finishing school?

Presentations of the products designed by the companies, technological processes, working environment, materials and business give a real picture of companies' work and products. To validate whether the students have understood the topic, the following questions are asked: How is the product made? What kind of technologies are used? What equipment and tools are necessary? Where is this company located? How rich are the companies and their owners? Is it produced by talented hand or advanced technology?

There are several ways to present pictures for a group of students. The most popular way is to do PowerPoint presentation. The authors also recommend using social network Pinterest.com to discover hundreds of related pictures, styles, inspiration, and many other ideas. This method gives schoolchildren the possibility to navigate and search for the pictures of their preference, which is more effective for the evocation phase.

2.4.2. Comprehension/Building knowledge phase – Technology cards

The comprehension and building of knowledge phase can be tightly related to the introduction and analysis of the technological processes used for the production of a particular product from provided materials. Introduction of technological processes could be started at school by choosing particular products and analysing the technological processes used for their production. It is necessary to involve subject teachers for a better explanation. There is a wide range of different informative materials in the Technology cards that can be used for this purpose (such as pictures and videos, introduction and information on the production processes, materials and equipments used).

Visits to the companies could be organized as the next step where students can see the technological processes with their own eyes, they would have better understanding of the specificity of the technological and work processes in this sector, also aspects, which can limit the observation (e.g., safety concerns in the heavy industries like metal and machinery industry; also noise, smells and heat which play role in many metal production processes).

2.4.3. Reflection/Consolidation phase

After visiting a company, the consultant asks questions such as:

- Is it worthwhile to become an expert in this field?
- What would be the benefits for me entering this industry?

Based on the analysis of the results performed during the project, the most important competencies in the field of economics evaluated by Baltic companies are:

- handling problematic situations adequately and on time, making necessary decisions;
- setting objectives for achieving specific aims in order to produce value-added products;
- assessing service cost calculations, required investments and workforce;
- project design and management.

Metal processing	TECHNOLOGY GROUP	**Metal cutting**	SPECIFIC TECHNOLOGY	**Turning, milling, drilling, laser cutting**

INTRODUCTION

Metal cutting process is an industrial process in which metal parts are shaped by removal of unwanted material. In traditional chip-forming processes, such as turning, drilling, and milling metal is removed as a plastically deformed chip of required dimensions.
Metal cutting is one of the most widely used methods to produce metal parts allowing to get parts with defined dimensions and shapes and required surface roughness. Sheet metal cutting is mostly used with the help of lasers or water.

KEY WORDS, ABBREVIATIONS

CNC – computer numerical control. Tool bit – a non-rotary cutting tool. Milling cutter – a rotary cutting tool.
Cutting fluid – fluid for cooling and lubrication during the metal cutting.
Lasers – a narrow single colour beam of light. Abrasive jet – jet of fine abrasive particles, usually about 0.025 mm in diameter.

Tool bit

Milling cutter

Drill

Cutting fluid

PROCESS DESCRIPTION

Turning is a machining process in which a non-rotary tool bit describes a helix toolpath by moving more or less linearly while the workpiece rotates. Turning can be done manually, in a traditional form of lathe, which frequently requires continuous supervision by the operator, or by using an automatic lathe, which does not. Today the most common type of such automation is CNC.
https://www.youtube.com/watch?v=8EsAxOnzEms
Milling is a cutting process that uses a milling cutter to remove material from the surface of a workpiece. The milling cutter is a rotary cutting tool, often with multiple cutting points. As opposed to drilling, where the tool is advanced along its rotation axis, the cutter in milling is usually moved perpendicular to its axis.
https://www.youtube.com/watch?v=Ef59DogwLrI
Drilling is a cutting process that uses a drill bit to cut a hole of circular cross-section in solid materials. The drill bit is usually a rotary cutting tool. The bit is pressed against the workpiece and rotated at rates from hundreds to thousands of revolutions per minute. This forces the cutting edge against the workpiece, cutting off chips from the hole as it is drilled.
https://www.youtube.com/watch?v=KYfAjakKO5w
Laser cutting Laser cutting works by directing the output of a high-power laser most commonly through optics. The laser optics and CNC are used to direct the material or the laser beam generated. The focused laser beam is directed at the material, which then either melts, burns vaporizes away, or is blown away by a jet of gas, leaving an edge with a high-quality surface finish.
https://www.youtube.com/watch?v=PlF_oXvbu4s

ECONOMIC FACTS AND DATA

Turning machine price	Drilling/Milling machine price	Laser metal cutting machine price
1,000–40,000 EUR … > 100,000 EUR	600–20,000 EUR … > 100,000 EUR	> 8,000 EUR … > 100,000 EUR

Price of the metal depends on the metal type, size and quality.
Low-carbon steel, sheet metal and standard profiles: 1.5–2.5 EUR/kg.
Stainless steel, sheet metal: 3.5–6.5 EUR/kg.
Aluminium alloys, sheet metal and standard profiles: > 3.5 EUR/kg.
Laser and water cutting: different sizes (e.g., sheets 4000×2000×20 mm).
Price of the instrument depends on the design and application.
Turning: tool holders 300–500 EUR/ps, inserts 10–20 EUR/ps.
Milling: tool holders 300–3000 EUR/ps, inserts 10–20 EUR/ps.
Drilling: solid carbide drills 40–300 EUR/ps.
Price of the metal cutting service depends on the technology and equipment.
Turning (CNC): 35–50 EUR/h.
Milling (hand/CNC): 25–35 EUR/h.
Drilling: 15–20 EUR/h.
Laser cutting: 25–35 EUR/h.

Fig. 5. Metal processing Technology card.

Design selection:

1.Product parts definition.

Size of product 200X200 mm

2.What pats may be produced/purchashed:

Art frame ❶ - produced,

Clock mechanism ❷ - purchashed

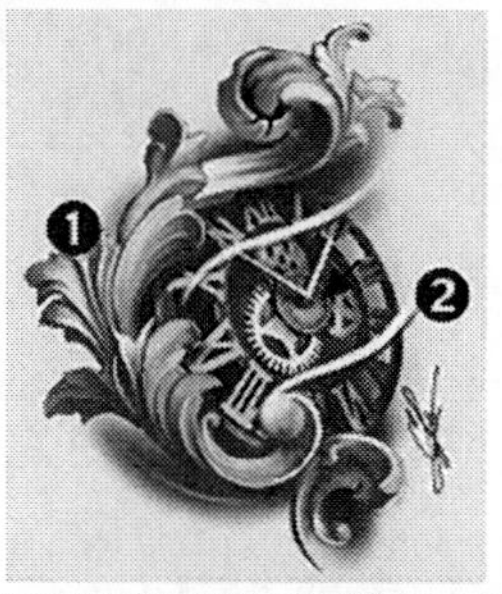

Technoloqy selection based on material:

1.Metal	2.Wood	3.Plastick
3D printing *Casting* *Stamping* *CNC milling*	*Hand cutting* *CNC cutting*	*3D printing* *CNC machining* *Injection molding*

Production type (PT):
Unit production (U); Batch production (B); Mass production (M)

Cost definition:

Material	*Machining type*	*PT*	*Qnt*	*1 part material cost, 1pcs/€*	*Preparation time, hours*	*Preparation cost, h/€*	*Processing time, hours*	*Processing cost, h/€*	*Comments*	*1 pcs direct cost, €*
Metal										
Copper	3D	U	1	200	3	20	4	50	Programming cost 60 €	460
Copper	Casting	B	6	23	1	0.5	0.5	10	Mould cost 1000 €	190
Copper	CNC machining	B	5	40	3,25	20	1	40	Programming cost 60 € + machine setup 5 €	141
Titanium	3D	U	1	200	3	20	4	50	Programming cost 60 €	460
	Casting	B								200
Stainless steel	CNC machining	B	5	10	3,25	20	1	40	Programming cost 60 € + machine setup 5 €	111
Stainless steel	3D	U	1	10	3	20	4	50	Programming cost 60 €	270
...										
Wood										
Wood	Handmade	U	1				25	20		500
Wood	CNC machining	B			3				Programming cost 60 €	110
...										
Plastic										
Plastic	3D	U	1	5	3	20			Programming cost 60 €	42
Plastic	Stamping	M							Stamp cost 10000	5
...										

Fig. 6. Example of practical work.

Career counsellors organize practical work for schoolchildren using product cost calculations in Technology cards, information gathered during visits to the companies, and also other information available on the Internet. Independent tasks for students can be related to business sustainability in the selected industry.

3. CONCLUSIONS

As a result of this project there will be teachers and schoolmasters having competencies in career education and guidance, implementing new interdisciplinary methods at secondary schools. The possibility to communicate with enterprises will give schoolchildren realistic impression of a selected profession.

Authors have prepared teaching and training materials for the secondary school teachers in the format of service training course “Innovative methods for implementing interdisciplinarity in career counselling (IMATEII)”. The knowledge and supervision of the researches and lecturers of technical universities from three Baltic countries will help young people to find the right way in their selection of profession.

ACKNOWLEDGEMENTS

This research was supported by Erasmus + project VERT17065 – Innovative methods for implementing interdisciplinarity in career counselling (IMATEII) and F15027 – Smart manufacturing and materials technologies competence centre (IMECC). The publication costs of this article were covered by the Estonian Academy of Sciences.

REFERENCES

1. 21st Century Skills for Students and Teachers. https://ainamulyana.blogspot.com/2017/06/21st-century-skills-for-students-and.html (accessed 2019-08-18).
2. Kenneth, B. H. Career education and education reform: time for a rebirth. *Phi Delta Kappan Int.*, 2001, **83**(4), 327–331.
3. Shevtshenko, E., Karaulova, T., Igavens, M., Strods, G., Tandzegolskienė, I., Tūtlys, V., Seyed, T., and Kuts, V. Dissemination of Engineering Education at Schools and its Adjustment to Needs of Enterprises. In *Proceedings of the 28th DAAAM International Symposium, Zadar, Croatia, November 08–11, 2017* (Katalinic, B., ed.). DAAAM International, Vienna, 2017, 44–53.
4. Tandzegolskienė, I., Tūtlys, V., Jurgilė, V., Strods, G., Igavens, M., and Shevtshenko, E. Attracting youth to the occupations in the food industry, agriculture and engineering: issues for policy and practice. In *Proceedings of the 11th annual International Conference of Education, Research and Innovation, Seville, Spain, November 12–14, 2018*. ICERI2018 Proceedings, 2018, 1637–1643.
5. Shevtshenko, E., Karaulova, T., Igavens, M., Strods, G., Tandzegolskienė, I., Tūtlys, V., Seyed, T., and Kuts, V. Implementing interdisciplinarity in career guidance for secondary school students in forestry and wood, metal and machinery, agriculture and food sectors of industry. *Holistic Learn.*, 2017, **3**, 53–60.
6. Crawford, A., Mathews, S., Makinster, J., and Saul, E. W. *Teaching and learning strategies for the thinking classroom*. The International Debate Education Association, New York, 2005.
7. Jonassen, D. H. and Rohrer-Murphy, L. Activity theory as a framework for designing constructivist learning environments. *Educ. Technol. Res. Dev.*, 1999, **47**(1), 61–79.
8. Meredith, K. S., Steele, J. L., and Kikusova, S. Critical Issues: democracy, community, self, literacy, and the value of global conversation. *J. Literacy Res.*, 2001, **33**(1), 169–202.
9. Fleming, G. Introduction to critical thinking. https://www.thoughtco.com/introduction-to-critical-thinking-1857079 (accessed 2018-09-24).
10. Kaganski, S., Majak, J., Karjust, K., and Toompalu, S. Implementation of key performance indicators selection model as part of the Enterprise Analysis Model. *Procedia CIRP*, 2017, **63**, 283–288.
11. Paavel, M., Karjust, K., and Majak, J. Development of a product lifecycle management model based on the fuzzy analytic hierarchy process. *Proc. Est. Acad. Sci.*, 2017, **66**(3), 279–286.
12. Maleki, M., Shevtshenko, E., and Cruz-Machado, V. Comparative Analysis of Customer Value Dimensions. *Eng. Econ.*, 2013, **24**(5), 488–495.
13. Kangilaski, T., Polyantchikov, I., and Shevtshenko, E. Partner network and its process management. In *Proceedings of ICINCO 2013*, 519–527.
14. Sahno, J. and Shevtshenko, E. (2014). Quality Improvement Methodologies for Continuous Improvement of Production Processes and Product Quality and Their Evolution. In *Proceedings of the 9th International DAAAM Baltic, Industrial Engineering, Tallinn, Estonia, April 24–26, 2014*, 181–186.
15. Kuts, V., Otto, T., Caldarola, E., Modoni, G., Sacco, M. Enabling the teaching factory leveraging a virtual reality system based on the Digital Twin. In *Proceedings of the 15th Annual EuroVR Conference, London, UK, October 22–23, 2018*. VTT Technical Research Centre of Finland, 2018, 26–31.

Innovaatilised meetodid inseneriõppe populariseerimiseks koolides

Eduard Shevtshenko, Tatjana Karaulova, Maris Igavens, Gunars Strods, Ilona Tandzegolskiene, Vidmantas Tutlys ja Kashif Mahmood

Viimastel aastatel on kõrghariduse sektoris olnud tehnilistel erialadel tegemist pideva tudengite nappusega. Selle probleemi lahendamiseks otsustasid kolm Baltimaade ülikooli alustada koostööprojektiga "Innovaatilised meetodid ja õpetamise tehnoloogiad interdistsiplinaarsuse juurutamiseks karjääri nõustamises". Projekti eesmärk on arendada uut lähenemist, mis toetab koolinoorte karjäärivalikut ülikoolide, üldhariduskoolide ja ettevõtete koostöö tulemusena. Autorid tutvustavad probleemi lahendamist metalli- ja masinatööstuse baasil, kuid antud projekti analüüsi tulemusi võib rakendada ka teistes tööstusvaldkondades.

Proceedings of the Estonian Academy of Sciences,
2019, **68**, 4, 364–370
https://doi.org/10.3176/proc.2019.4.03
Available online at www.eap.ee/proceedings

MANUFACTURING
ENGINEERING

Synchronizing physical factory and its digital twin through an IIoT middleware: a case study

Vladimir Kuts[a*], Gianfranco E. Modoni[b], Tauno Otto[a], Marco Sacco[c], Toivo Tähemaa[a], Yevhen Bondarenko[a], and Ruxin Wang[a]

[a] Department of Mechanical and Industrial Engineering, School of Engineering, Tallinn University of Technology, Ehitajate tee 5, 19086 Tallinn, Estonia
[b] Institute of Intelligent Industrial Technologies and Systems for Advanced Manufacturing – National Research Council of Italy, Giovanni Amendola 118, 70126 Bari, Italy
[c] Institute of Intelligent Industrial Technologies and Systems for Advanced Manufacturing – National Research Council of Italy, Via Alfonso Corti 12, 20133 Milano, Italy

Received 26 March 2019, accepted 13 May 2019, available online 22 October 2019

Abstract. Digital twin (DT) is the virtual clone of a factory representing its static and dynamic aspects (e.g., processes, systems, products, etc.) in detail. Among the significant challenges that manufacturing company has to face to implement the DT, one of the most demanding is applying an appropriate software infrastructure, which would enable synchronization of the physical factory with its DT. In this case, it was possible to exploit wide range of capabilities of DT in its full potential. In particular, the DT was used in different conditions to enable various operations within the shop floor, to simulate and assess the factory's performance. To support companies in addressing this challenge, this paper presents a potential solution, based on the Industrial Internet of Things (IIoT) middleware, that implements a fully dual-way synchronization between the real and virtual worlds. A case study was carried out to investigate the possibilities to implement the solution. To demonstrate correctness and validity of the approach, tests were carried out in the laboratories of Flexible Manufacturing Systems, Robotics Demo Centre and ProtoLab of Tallinn University of Technology (TalTech).

Key words: digital twin, IIoT, virtual reality, industrial robots.

1. INTRODUCTION

One of the key aspects for the success of a modern factory is realizing the importance of using its digital twin (DT), which is a virtual clone of factory's manufacturing system (or part of it) that represents its static and dynamic aspects in detail [1]. The creation of this virtual representation can be exploited to enable various strategic operations within the shop floor, such as simulating and assessing the factory's performance. For example, the simulated environment provided by the DT can allow production managers and the integrators of factory's automation system to check the changes applied to the factory layout without the need to test them in the real production environment. Also, opportunities of virtual reality are used to enhance the environment. This allows users to create the envisioned factory in virtual reality (VR) and to inspect manufacturing systems in the real scale.

* Corresponding author, vladimir.kuts@taltech.ee

Authors of this research focus on a specific application of the DT – on management and control of the manufacturing systems, mainly in the field of industrial robots. In particular, the idea behind the work is that human presence can be simulated, accompanied, and enhanced by other operators remotely (multiplayer capability) by using VR tools. In previous studies of the robotics, virtual commissioning (VC) has mainly been used to create a simulation to optimize robots' algorithms during development phase. In more general manufacturing field, the simulations have been exploited to represent the complex behaviour of a system, also taking the possible consequences deriving from external factors like human interactions or design constraints into consideration. The potential of the DT has been recently analysed in various publications [1–4]. Schluse and Rossmann [2] described an idea of DT as "Versatile Simulation Database" (VSD), which is a static data container containing related algorithms/interfaces to manipulate data. Moreover, it also provides functionalities for parallel and distributed simulation. For a specific case, Grinshpun et al. [3] illustrated an application involving KUKA LWR4 robot using VEROSIM as a simulation environment. VEROSIM provides a scripting language called SOML++, where engineers can easily monitor occurring force and torque during the manufacturing simulation. Schroeder et al. [4] see DT as a composition of different models and data, and that it can be aggregated into an AutomationML model enhancing the exchange data between heterogeneous systems via a middleware named FIWARE [5].

However, according to numerous works found in the literature [6–8], a realization of the DT requires the following key features:

(a) the data acquisition via a monitoring system;
(b) setup of the detailed synchronization of the DT in order to make achievement of specific goals successful;
(c) using a data model to represent the real factory and its performance;
(d) synchronization between the real and digital factory.

This paper focuses mainly on the last feature, proposing the idea that middleware is used for synchronization of the robot and its DT, which propagates the data stream of the factory telemetry from the physical factory assets towards their digital representation and back. The presented work has also its roots in previous works from the same authors concerning an infrastructure architecture for the synchronization [7,8], and the industrial robot tests combining VR and DT [8–10]. Also, this work takes other studies in similar fields and comprising remote laboratories [11], initial design [12] and virtual training [13] as a reference. The remainder of this paper is structured as follows: Section 2 presents the software infrastructure, while Section 3 illustrates its application and evaluation within a case study. Section 4 presents the benefits of an approach based on a synchronized DT. Finally, Section 5 concludes, summarizing the major findings.

2. SOFTWARE INFRASTRUCTURE FOR THE BIDIRECTIONAL SYNCHRONIZATION OF REAL AND VIRTUAL ASSETS

Modern manufacturing systems are projected into a new era in which they will be capable of exploiting the IIoT protocol in its full potential. Leveraging this potential, manufacturing systems will become dynamic networks comprising of different kinds of physical and virtual resources. In particular, thanks to the advanced communication and networking capabilities provided by the IIoT, each production component connected to the IIoT network will be paired with its fully synchronized DT. To contribute realizing this vision, present work envisions a software infrastructure to constantly synchronize the physical components with their digital representations. This infrastructure leverages a message-oriented Industrial Internet of Things (IIoT) middleware, based on a publish-subscribe interaction [14]. The middleware handles the factory telemetry, i.e., the real-time data stream, which involves all data such as monitored controlled variables coming from the plant and actuator directives to be promptly executed on a physical resource. Thus, the middleware allows any (real or virtual) components connected to the IIoT protocol to propagate significant information to keep other interested components (information consumers) updated about occurred events (e.g., the position change of a device). Middleware can filter the content of each message and dispatches the information updates only to interested resources, where information dispatching adopts an event-based PUSH strategy, thus announcing the information updates under the form of events. Thanks to the middleware, notifications can be distributed as messages, so that the information consumers can take further evaluations and then perform some needed actions. Also, the middleware allows the interested components to specify their requests to be subscribed to the updates related to the specific contextual factory information. Specifically, each controller of the physical component (*PhysicalResourceController*), after having acquired and collected the monitored information, can share the context information with other components, and in particular with the component that manages the DT (*DigitalTwinController*). In turn, the latter can send directives to the *PhysicalResourceController*s according to the input received from the virtual environment.

Under these conditions, *PhysicalResourceControllers* and *DigitalTwinController* play the role of prosumers, i.e., they can be both, information producers and consumers.

3. THE CASE STUDY

A proof of concept of the software infrastructure to synchronize a real robot with its DT was implemented in the Flexible Manufacturing Systems (FMS), Robotics Demo Centre and ProtoLab of TalTech. The latter includes the Industrial Robotics' section, which consists of two heavy load robots. One is Yaskawa Motoman GP8 with a changeable tool for picking and placing tasks. The herein proposed infrastructure was built focusing on this robot. In this regard, its virtual model was created through the Unity3D game engine platform. The visualization of this model requires using standard VR equipment (e.g., head-mounted display units, such as Oculus Rift/HTC Vive).

For the re-creation of a real Industrial Robot, a cell was created with the aid of 3D modelling software, producing a virtual representation of the laboratory with all its assets (Fig. 1). The more precise process of creation of digital model is described in previous research papers of the authors with all aspects of the technical side [9]. A short list of most essential aspects, which should be considered while digitalizing manufacturing equipment, using own 3D models based on drawings or imported from the libraries of equipment manufacturer, is the following:

- rigging of the model – the central pivot points for movable objects should be defined appropriately;
- rendering time – the 3D objects, geometry should be simplified to improve the performance of the visual simulation, the mesh should be correct;
- re-checking and keeping one-to-one scale for the accuracy;
- proper kinematics of portable equipment.

If the aspects of this list are not taken into consideration, the accuracy of a DT will not be efficient for the experiments.

Figure 2 shows five different components of the software infrastructure:

(a) *RobotController*, which controls the robot through a specific API;
(b) *DigitalTwinController*, which manages the DT model;
(c) a publish-subscribe middleware as introduced in the previous section;
(d) *RobotCommunicator*, which handles the communication for the real component;
(e) *DigitalTwinCommunicator*, which handles the communication for the virtual component.

Fig. 1. Digitalized laboratories – Flexible Manufacturing Systems, Robotics Demo Centre, and ProtoLab (a, b, c – real; d, e, f – replica).

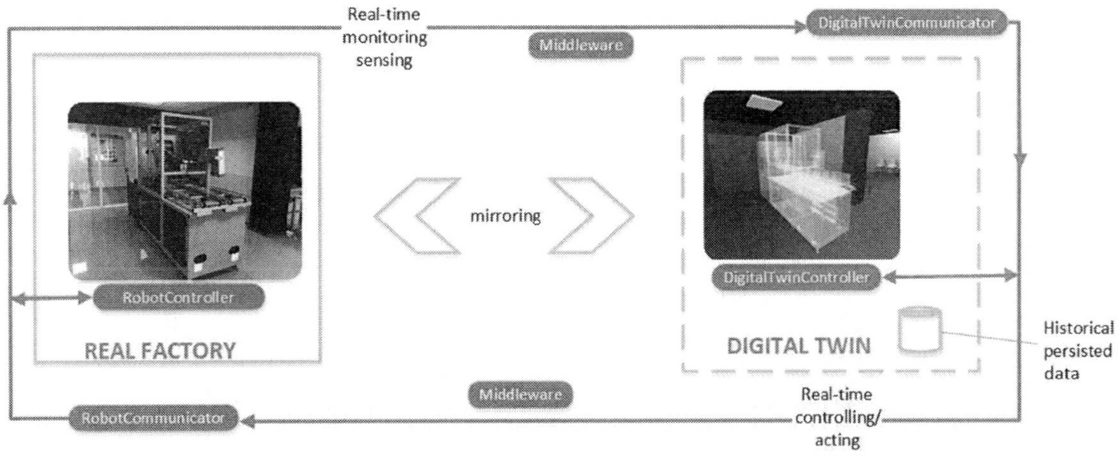

Fig. 2. Synchronization process through the proposed software infrastructure.

Also, Fig. 1 represents how these components are linked to enable the synchronization process, which in particular, is enabled by the middleware and two of its clients (*RobotCommunicator* and *DigitalTwinCommunicator*). The role of middleware is in turn played by Message Queuing Telemetry Transport (MQTT) publisher/subscriber protocol, which is standardized by ISO [15] and known for being a lightweight and fast solution for IoT applications.

Another potential option for middleware could have been Kafka platform [16], which allows greater scalability and data persistence. However, MQTT was adopted for its ease of implementation and shorter setup time, while also providing the needed features for the considered case study. Also, current MQTT implementations provide countermeasures against various failures or error situations by offering features, which make middleware more resilient.

All messages published by the MQTT clients are sent to the MQTT broker, which then forwards the messages to other clients subscribed to the same messaging topic (the two *Monitors* are at the same time data publisher and consumer). The presented solution utilizes six message topic pairs, which correspond to the control inputs and monitoring outputs of the *RobotListener* (i.e., setting and monitoring of the robot position). The MQTT broker was hosted on a public IP server, managed by Digital Ocean [17], and located in Frankfurt, Germany. The real robot can be managed by running *DigitalTwinCommunicator* script, which allows the robot to be connected using MQTT.

Another challenge was related to Unity. Indeed, since it does not have MQTT integration out of the box, it was necessary to implement a custom solution, which would allow starting MQTT client scripts directly from Unity runtime. However, MQTT protocol is universal, and the clients were already implementing it in many programming languages, including C# used by Unity. The final solution for this demo was based on .NET MQTT client implementation by Eclipse [18] and existing test Unity project available on GitHub [19].

An important reason for choosing a Unity3D game engine for development, was possibility to build both, VR-supported and classic desktop versions of the application. Though both versions feature the same DT control interface, VR version was also complemented with user avatar networking, allowing researchers to test the robot online while being together in the same virtual environment (Fig. 3). This feature allows demonstrating an important aspect of the factories connected in the future – the ability to monitor equipment and collaborate in real-time, even from remote locations.

The aspects that can affect the success for such a real-time, multi-user, internet-based collaborative VR systems, are the latency, i.e., the time needed to synchronize the virtual world with the real one and vice versa, the scalability and user experience. These aspects depend in turn on the number of the connected IIoT devices, the throughput of the information shared between the devices and the size of the 3D scene (nodes, geometries, assets). Thus, after the implementation, the solution was evaluated according to the following three criteria:

(a) latency (the aim of the test is to measure it and demonstrate that it is under a fixed threshold);
(b) capability to measure according to the throughput, which can be increased by changing the status update frequency of the robot;
(c) capability to control DT from remote locations.

To perform the above mentioned experiment, a test was performed by changing two parameters: robot send

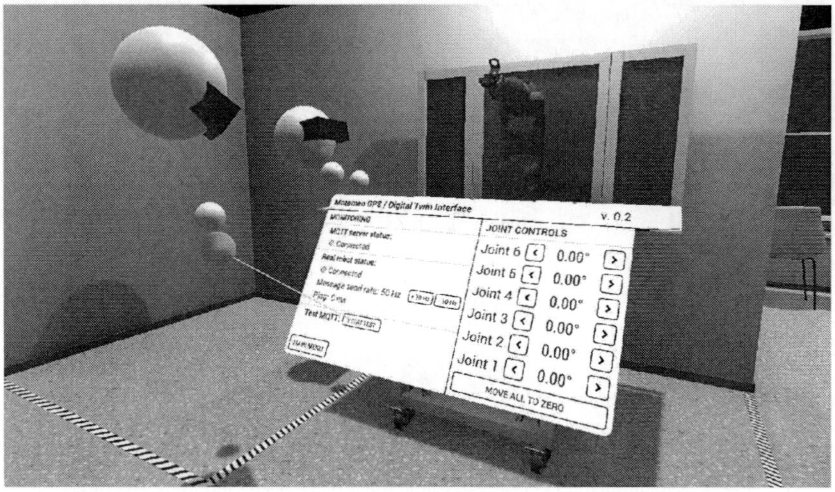

Fig. 3. Researchers' avatars controlling the robot in VR.

rate (how often *RobotListener* sends updates about the robot's position to the broker) and MQTT Quality of Service (QoS) level (it determines how reliably the messages are delivered by the MQTT protocol). Send rate values were taken in the range from 10–100 Hz, with a step of 10 Hz, while QoS level can take values of 0, 1, or 2. Greater values require more computation time to process messages. Finally, the time required for sending messages from *DigitalTwinCommunicator* to *RobotCommunicator* and back (ping), was taken as a metric of connection speed. For each combination of parameters, the robot was sent 50 movement commands and ping was continuously sampled, to calculate the average for each experiment. The test was run in two different places at premises of STIIMA-CNR (Bari, Italy) and of TalTech Industrial Virtual and Augmented Reality Laboratory (IVAR Lab), which is part of Flexible Manufacturing Systems and Robotics Demo Centre. Data collected after running tests in these two locations can be seen in Fig. 4.

A significant result obtained from the experiment is the fact that different values of the send rate used in the experiment do not affect the time of synchronization. The only parameter affecting the communication speed is QoS – rising level to 2 results in longer ping times (as the message processing time increases), while level 0 and 1 provide approximately equal results. This proves that it is possible to maintain a fast and reliable connection even with high send rate and an acceptable QoS level.

4. BENEFITS OF THE PROPOSED SOLUTION AND FUTURE DEVELOPMENTS

A list of benefits deriving from the exploitation of the proposed infrastructure is the following:

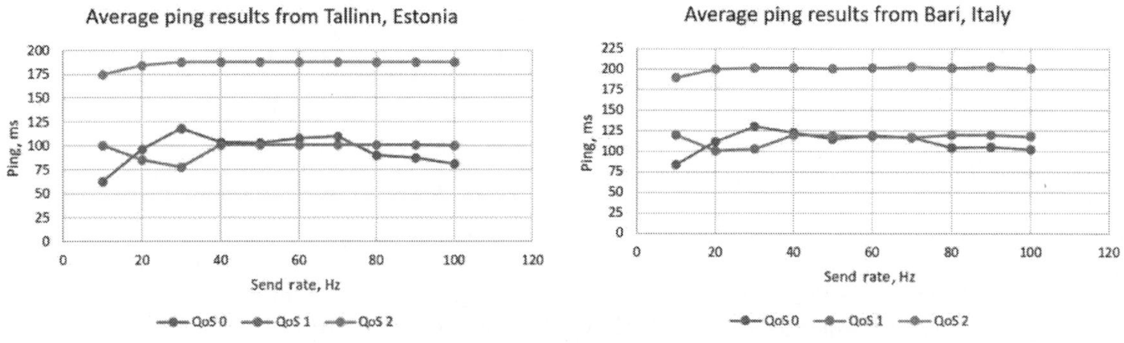

Fig. 4. Experiment results.

- it is possible to work with the system remotely in real time, eliminating the need to be in close contact with dangerous equipment or hazardous environments.
- support of multiplayer mode and remote collaboration. IIoT-enabled devices can efficiently share their work with many clients, allowing them to create a single networked virtual environment where workers can directly communicate while managing the connected equipment. DT can support training with the equipment virtually as training in reality may not be feasible (e.g., too expensive, etc.).
- possibility to create simulations, which allow to "rewind" the state of the virtual system to an arbitrary point of time to identify failures or analyse system's actions, exploiting both, real data coming from the robot and historical data previously collected; and then persisted on the database (a feature to be implemented). Moreover, these capabilities can allow creating virtual sensors, which can send commands to the real machines in order to evaluate performance and track possible collisions with personnel or solid bodies.
- DT brings production efficiency to a new level. Indeed, instead of taking down the manufacturing line to apply some changes, everything can be tested and verified on the virtual clone, and then safely put into production. Human-robot interactions can be tested in the virtual environment in offline mode. In addition, industrial equipment does not need to be stopped while re-programming, which leads to time-saving.

In the future, speed optimization has to be developed to synchronize and integrate other assets of different new and existing labs with the middleware. In addition, the potential of Kafka in the role of middleware has to be investigated. Moreover, the DT will be paired with the database to maintain the history of robot's positions, which can then be used to replay past robot movements (See Fig. 1) [20]. Finally, the overall infrastructure will be tested in other fields also different from manufacturing (e.g., telemedicine).

5. CONCLUSIONS

Results of this work describe implementation of IIoT middleware, which supports full dual-way synchronization between a real robot and its virtual counterpart. An experimental application showed efficient connectivity, and the researchers from STIIMA-CNR (Bari, Italy) were able to control and manage industrial robot situated in TalTech IVAR Lab (Tallinn, Estonia). Moreover, the work gave the opportunity to make an international collaboration, creating the basis for the integration of different laboratories and factory environments into a digital network.

ACKNOWLEDGEMENTS

This research was supported by the project AR16077 Smart Industry Centre (SmartIC) No. 2014-2020.4.01.16-0183, supported by the EU Regional Development Fund. The authors are also grateful to the Computer and Systems Engineering student of the Tallinn University of Technology – Ninel Dudnyk, for her help in the experiments with laboratory digitalization. The publication costs of this article were covered by the Estonian Academy of Sciences.

REFERENCES

1. Modoni, G. E., Caldarola, E. G., Sacco, M., and Terkaj, W. Synchronizing physical and digital factory: benefits and technical challenges. *Procedia CIRP*, 2019, **79**, 472–477.
2. Schluse, M. and Rossmann, J. From simulation to experimentable digital twins: Simulation-based development and operation of complex technical systems. In *Proceedings of 2016 IEEE International Symposium on Systems Engineering (ISSE), Edinburgh, UK, October 3–5, 2016.* https://ieeexplore.ieee.org/abstract/document/7753162
3. Grinshpun, G., Cichon, T., Dipika, D., and Rossmann, J. From Virtual Testbeds to Real Lightweight Robots: Development and deployment of control algorithms for soft robots, with particular reference to. In *Proceedings of ISR 2016: 47st International Symposium on Robotics, Munich, Germany, June 21–22, 2016.* https://ieeexplore.ieee.org/document/7559118/authors#authors
4. Schroeder, G. N., Steinmetz, C., Pereira, C. E., and Espindola, D. B. Digital Twin Data Modeling with AutomationML and a Communication Methodology for Data Exchange. *IFAC-PapersOnLine*, 2016, **49**(30), 12–17.
5. FIWARE. https://www.fiware.org/
6. Modoni, G. E., Sacco, M., and Terkaj, W. A Telemetry-driven approach to simulate data-intensive manufacturing processes. *Procedia CIRP*, **57**, 281–285.
7. Kuts, V., Modoni, G. E., Terkaj, W., Tähemaa, T., Sacco, M., and Otto, T. Exploiting factory telemetry to support Virtual Reality simulation in robotics cell. In *Augmented Reality, Virtual Reality, and Computer Graphics. Proceedings of the 4th International Conference, AVR 2017, Part 1, Ugento, Italy, June 12–15, 2017* (De Paolis, L. T., Bourdot, P., and Mongelli, A., eds). Springer International Publishing, 2017, 212–221.
8. Kuts, V., Sarkans, M., Otto, T., and Tähemaa, T. Collaborative work between human and industrial robot in manufacturing by advanced safety monitoring system. In *Proceedings of the 28th International DAAAM*

Symposium Intelligent Manufacturing and Automation, Zadar, Croatia, November 08–11, 2017 (Katalinic, B. ed.). DAAAM International, Vienna, Austria, 2017, 0996–1001.

9. Kuts, V., Otto, T., Tähemaa, T., and Bondarenko, Y. Digital Twin based synchronised control and simulation of the industrial robotic cell using Virtual Reality. *J. Mach. Eng.*, 2019, **19**(1), 128–145.
10. Kuts, V., Otto, T., Tähemaa, T., Bukhari, K., and Pataraia, T. Adaptive industrial robots using machine vision. In *Proceedings of the ASME 2018 International Mechanical Engineering Congress and Exposition, Pittsburgh, Pennsylvania, USA, November 9–15, 2018.* https://asmedigitalcollection.asme.org/IMECE/proceedings-abstract/IMECE2018/52019/V002T02A093/276346 (accessed 2019-01-15).
11. Ji, W., Yin, S., and Wang, L. A Virtual training based programming-free automatic assembly approach for future Industry. *IEEE Access*, 2018, **6**, 43865–43873.
12. Sell, R. Remote laboratory portal for robotic and embedded system experiments. *Int. J. Online Eng.*, 2013, **9**(8), 23–26.
13. Sell, R., Coatanéa, E., and Christophe, F. Important aspects of early design in mechatronic. In *Proceedings of the 6th International DAAAM Baltic Conference INDUSTRIAL ENGINEERING, Tallinn, Estonia, April 24–26, 2008* (Küttner, R., ed.). http://innomet.ttu.ee/daaam08/Online/Design%20Engineering/Coatanea-Sell.pdf
14. Modoni, G. E., Veniero, M., Trombetta, A., Sacco, M., and Clemente, S. Semantic based events signaling for AAL systems. *J. Ambient Intell. Hum. Comput.*, 2018, **9**(5), 1311–1325.
15. ISO/IEC 20922:2016 Information Technology – Message Queuing Telemetry Transport (MQTT) v3.1.1. https://www.iso.org/standard/69466.html
16. Apache Kafka at GitHub. https://github.com/apache/kafka (accessed 2018-12-12).
17. Digital Ocean cloud hosting platform. https://www.digitalocean.com/ products/spaces/ (accessed 2018-12-17).
18. Eclipse Paho MQTT library for .NET at GitHub. https://github.com/eclipse/paho.mqtt.m2mqtt (accessed 2018-12-16).
19. M2MqttUnity at GitHub. https://github.com/gpvigano/M2MqttUnity (accessed 2018-12-16).
20. Modoni, G. E., Doukas, M., Terkaj, W., Sacco, M., and Mourtzis, D. Enhancing factory data integration through the development of an ontology: from the reference models reuse to the semantic conversion of the legacy models. *Int. J. Comput. Integr. Manuf.*, 2017, **30**(10), 1043–1059.

Füüsiline tehas ja selle digitaalse kaksiku sünkroonimine tööstusliku nutistu vahevara kaudu: juhtumi uuring

Vladimir Kuts, Gianfranco E. Modoni, Tauno Otto, Marco Sacco, Toivo Tähemaa, Yevhen Bondarenko ja Ruxin Wang

Digitaalne kaksik (DT) on tehase või tehnoloogiaseadme virtuaalne kloon, mis võimaldab erineva detailsusega ka selle staatilisi ja dünaamilisi vaateid (näiteks protsessidest, süsteemidest, toodetest jne). Olulisemate väljakutsete hulgas tootmisettevõttele digitaalse kaksiku metoodika rakendamisel on nutika tootmise taristu väljaarendamine, mis võimaldaks füüsilist tehast digitaalse kaksikuga sünkroonida. Digitaalseid kaksikuid võib seejuures kasutada tootmisoperatsioonide simuleerimiseks ja tehase jõudluse hindamiseks. Selle hõlbustamiseks on artiklis pakutud tööstusliku nutistu vahevaral põhinev lahendus, mis võimaldab kahesuunalist sünkroonimist reaalsete ja virtuaalsete maailmade vahel. Lisaks on analüüsitud lahenduse rakendusvõimalusi reaalse juhtumiuuringu raames, mis viidi läbi TalTechi paindtootmise ja robootika demokeskuses.

Proceedings of the Estonian Academy of Sciences,
2019, **68**, 4, 371–375
https://doi.org/10.3176/proc.2019.4.04
Available online at www.eap.ee/proceedings

MANUFACTURING
ENGINEERING

Optimization of enterprise analysis model for KPI selection

Sergei Kaganski*, Martin Eerme, and Ernst Tungel

Department of Industrial and Mechanical Engineering, Tallinn University of Technology, Ehitajate tee 5, 19086 Tallinn, Estonia

Received 12 April 2019, accepted 6 May 2019, available online 22 October 2019

Abstract. Nowadays, a number of methods and principles are available in literature for KPIs (key performance indicators) selection. However, they are facing with the same issue – lack of optimal procedures for selection of suitable metrics for particular company. The purpose of the study is to optimize enterprise analysis model (EAM) for KPIs selection and to reduce the time and resources necessary for the analysis of the enterprise. In the current study four outlier's detection methods for eliminating "outliers" in the answers are utilized. Furthermore, the experts from production and academic institutions are participating in evaluation and analysis of questionnaires. The optimized EAM is going to help simplifying the choice of KPIs, reducing the amount of data and optimizing the data flow. The optimized set of questions in EAM and KPIs that could be used in companies for improving their productivity are determined. The research is focused on SMEs (small and medium enterprises) and intention is to increase their competence on the market. The general procedure for KPIs selection/optimization for SME is pointed out.

Key words: key performance indicators (KPIs), KPIs selection model, enterprise analysis model (EAM), outlier's detection methods.

1. INTRODUCTION

The last two decades have seen profound change in the private sector's relationship with society and reconsideration of the links between state and market. The basic principles, on which private companies are expected to contribute to the public sector, are altered due to globalization accompanied and reinforced by social and economic reforms [1–3]. In comparison with other type of enterprises, the SMEs are showing good performance [4–7]. Nowadays, to be able to face all challenges and become winner in competition "be eaten or stay alive", companies need to deal within dynamic environment of fierce competition, shrinking budgets and heavy price pressures [8]. From one point of view, SMEs cannot effort high investments on research and development (R&D) like large enterprises or concerns, due to the financial aspects. From another point, the speed of the implementation of new technologies, approaches and methods are higher [9]. One of the methods that could help companies to improve the situation at production or eliminate difficulties, is the implementation and measurement of right KPIs.

It has been stated that companies which have developed and implemented sustainable practices, were able to improve product quality by following quality ratio, first pass yield and increased profits [10].

As a result, new trends/approaches that are becoming more popular, can be outlined – analytic hierarchy process (AHP) and SMART criteria synthesis for prioritizing company's KPIs, proposed by Shahin and Mahbod [11], where the pair-wise comparison has been done to judge the impact of the selected metrics on the company; the three components (academic, researching

* Corresponding author, sergei.kaganski@outlook.com

and supporting activities) that would simplify the importance of the metrics and AHP technic as selection methodology by Kadarsah [12]; Parmenter's 12-way model, which is based on four main stones, where the main focus is on the investigation and preliminary work [13]; questionnaire survey and confirmatory factor analysis (CFA) for data evaluation, where the consistency check of the data, received by analysing the questionnaires, has been performed by Yuan, Wang, Skibniewski and Li [14]; AHP based selection and prioritization methodology for leading KPIs by Podgorski [15].

2. KPI SELECTION MODEL

In [16] it is attempted to combine the advantages of the existing models and to avoid the disadvantages pointed out above. The proposed model can be divided into three phases:

(1) enterprise analysis model (EAM);
(2) data collection;
(3) data analysis and implementation.

The KPIs selection process is performed as follows:

- analysis of the enterprise. The EAM as a first phase of the selection is used to collect not only the general information about the company where study has been conducted, but also outline the weak spots faced by the management of the enterprise. The questionnaire is filled in by the employees, based on their position at the company (different amount and different questions). This allows to feature KPIs selection for particular company or type of companies, etc.;
- data collection for the analysis. A web-based questionnaire is composed as a rule – this enables to collect and analyse data more quickly and with higher efficiency (manual data collection is considered for special cases). The collected data are verified and stored in the database (server-cloud);
- data analysis. Sorting and grouping by numbers of respondents, applying weights. The answers are being analysed and evaluated by their importance (the answers are ranked by the 6-point scale, where 1 means that the answer is critical and needs to be taken into account and 6 means that the situation described in the question is not critical for the company);
- KPIs selection. The package of KPIs is selected by the expert group and two approaches: SMARTER criteria and fuzzy analytical hierarchy process (AHP);
- KPIs' implementation (the selected package of metrics is implemented by the company based on their ranks).

In addition, it is wise to mention that, independent from the used approach, the whole process of selecting KPIs is a continuous process. Management should not stop, when the package of metrics has been selected and implemented. The situation in the world is changing rapidly and new trends appear. We can see the same in production: new machines and robots, and the level of automatization is rising. There are new enterprise resource planning (ERP) systems, new manufacturing processes, new approaches etc. Due to these factors, the metrics that had to be followed in the beginning do not have to be analysed at present. As the process is continuous and the whole procedure of the KPIs' selection needs to be repeated by the management during a time frame, the optimization of the EAM is required.

3. ENTERPRISE ANALYSIS MODEL

The EAM is used to analyse the enterprise, to discover and to understand the critical spots. The following EAM goals can be outlined [17] as follows:

- getting the general information about the enterprise (field of action, number of employees, etc.);
- discovering the critical spots (based on the answers of the questions).
- providing the information which data should be collected based on the critical spots to eliminate the amount of unnecessary data (as there are links between KPIs and questions).

The proposed selection model of KPIs eliminates the main disadvantages of the previous concepts [11,12,18] as selection of the metrics is based on the issues discovered by the EAM after implementation in the studied company.

The model is based on traditional questionnaire for management and company's workers. The main goal is to bring out the bottlenecks and weaknesses in the production and general processes of the enterprise by analysing the answers. The questionnaire is one of the oldest and more commonly used tools for data collection. The advantages of this approach are availability (cheap and easy to establish) and quality of the answers (additional training is not necessary for the participants).

Simplified concept of the EAM has been illustrated in Fig. 1.

The questionnaire includes mapping the enterprise, there will be formed questions about certain fields of the company. General information regarding objectives, missions, visions, location of the enterprise etc. is received by the answers. More than 70 research papers with similar topics were studied to prepare this survey. Questions were divided into 15 categories, where

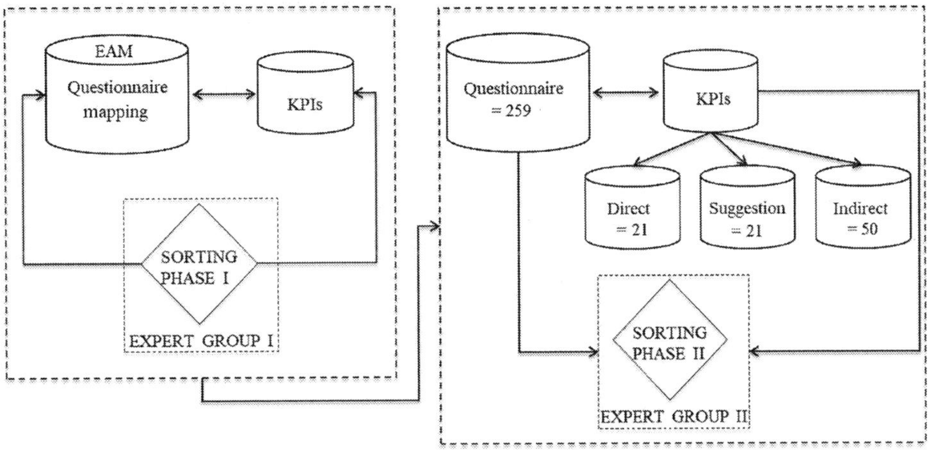

Fig. 1. The concept of the EAM.

each category had a different number of constructs. Constructs were linked to each group to simplify categorization. More than 40 different scales were used for the answers [19].

All questions have been composed based on the previous researches, which in case of the current study has higher impact on the final results due to the fact that questions describe different situations in the company. Furthermore, the questions or issues can be declared as common problems in the companies in general. By answering the questions, management will understand, what kind of problems they have in the company and based on that, what are the measures that have to be taken to change the situation in a positive way.

The KPIs for the study were selected after working through the analysis of different researches in that field (over 70 articles). The questions were linked with KPIs that would be worth investigation after data collection. In turn, the KPIs were divided into 3 groups to simplify linking:

(1) direct KPIs – indicators, which were in explicit correlation with the answers;
(2) indirect KPIs – indicators, which were connected with more than one question;
(3) suggested KPIs – indicators, which were proposed to the management for further study.

Indirect KPIs can be specified by answering at least two questions. For example, employee satisfaction can be followed only by knowing the staff turnover, availability of trainings and benefits etc.

To eliminate misunderstanding and provide better effectiveness of survey, logical connection between next pairs were established and tested:

- constructs + questions;
- questions + KPIs.

According to Fig. 1, sorting was performed in two phases by using web tool "Optimal Workshop". It enabled to save time not only on establishing the sorting process (input) but also on performing the evaluations (output). During the first stage, raw version of subjects was tested. The main task for the expert group was to match the left side (constructs/questions) with the right side (questions/KPIs). The case study group consisted of 10 researchers: 7 from different industrial companies, 2 from university and 1 from competence centre with no direct connection with the study, but with necessary knowledge and experience in that field. If questions or KPIs were matched with objects in a wrong way, the formulation of questions was changed or questions were replaced by more suitable ones.

The second phase was performed after improvements of the first phase. The main goal was to exclude further errors and to confirm, that previous review was done in a right way. The total number of questions, including corrective actions, was 259, which in turn were linked to 92 KPIs.

4. OPTIMIZATION OF EAM

The optimization of EAM is introduced based on workgroup's long term experience in area of linear and nonlinear constrained optimization, covering design of materials, structures and manufacturing processes [20–23]. The optimization is performed with an aim to develop an effective enterprise analysis model, which enables to perform analysis in a reasonable time frame without remarkable loss in quality. For this reason, first a thoroughgoing set of questions and KPIs was composed and then, this set was limited to bounds allowing resource effective analysis. The optimization problem considered can be formulated as follows:

$$
\begin{gathered}
Min\ R_i \\
\text{subjected to} \\
KPI \leq KPI^* \\
\text{questions} \leq \text{questions*}, \\
PI > PI^* . \qquad (1)
\end{gathered}
$$

In Eq. (1), R_i stands for the resources (total time for analysis, working hours for completing the questionnaire, etc.), KPI^* and questions* are estimated upper limits for the number of KPIs and questions, respectively. In order to keep model adequate, the information related to production PI should be retained upper critical limit PI^*. Optimization procedure was proposed in Eq. (1) to solve posed optimization problem.

In addition, during ranking phase, the ranked questions, received from experts, may contain outliers that have different impact on the data set when compared with others. Outliers may have critical impact on the data analysis. The goal was to optimize the questions by eliminating faulty answers from the total range [24]. Four different and simple outlier's detection methods were chosen:

(1) standard deviation method;
(2) modified Z-score method;
(3) Tukey's method;
(4) adjusted boxplot.

The basic steps of the proposed KPI selection/optimization procedure can be outlined as follows:

Step 1. Forming initial questionnaire, KPIs.
Step 1.1. Composing initial questionnaire based on literature, experts.
Step 1.2. Composing initial KPIs.
Step 1.3. Identifying links between constructs and questions.
Step 1.4. Identifying links between questions and KPIs.
Step 1.5. Classification of KPIs (direct, indirect, suggested).
Step 2. Applying an expert group to reduce questions, KPIs.
Step 2.1. Omitting questions unrelated or weakly related to KPIs.
Step 2.2. Omitting questions/KPIs, with no or weak impact on production.
Step 2.3. Ranking questions.
Step 3. Applying outlier's method for reducing questions, KPIs.
Step 3.1. Selection of outlier's methods.
Step 3.2. Employing the standard deviation method.
Step 3.3. Employing the Z-score method, modified Z-score method.
Step 3.4. Employing Tukey's method.
Step 3.5. Employing the adjusted boxplot method.
Step 3.6. Selection of outliers based on the results of applying outlier's methods.
Step 4. Estimating the final set of KPIs and questions.

Return back to Step 2 in case the number of questions and KPIs are still too huge to perform effectively in SME (KPI ≤ KPI*, questions ≤ questions*).

Note that contrary to the standard approach, in the case of the posed optimization problem, the initial solution is selected consciously infeasible. The first two constraints of Eq. (1) are not satisfied due to the thoroughgoing set of questions and KPIs considered as candidates for the final set.

5. CONCLUSIONS AND FURTHER STUDY

In the current study, the first phase (EAM) of KPIs selection model was described in detail. The EAM improvement problem was formulated as optimization problem. The optimization procedure, featured for particular problem, was developed. The outlier's detection methods were chosen for further study. Implementation of the optimization procedure allows to reduce time necessary for analysis of the enterprise. The selection model KPIs can be considered as foundation on which decisions and improvements would relay. Furthermore, it should simplify the work of management and make production more transparent. The future study is related to the application and refinement of the optimization procedure for KPIs' selection model.

ACKNOWLEDGEMENTS

This research was supported by the Innovative Manufacturing Engineering Systems Competence Centre (IMECC), supported by the Enterprise Estonia and co-financed by the European Union Regional Development Fund, project EU48685; Estonian Research Council (grant PUT1300); and the Estonian Centre of Excellence in Zero Energy and Resource Efficient Smart Buildings and Districts (ZEBE, grant TK146), funded by the European Regional Development Fund. The publication costs of this article were covered by the Estonian Academy of Sciences.

REFERENCES

1. Raynard, P. and Forstater, M. Implications for small and medium enterprises in developing countries. UNIDO and the World Summit on Sustainable Development, Vienna, 2002. https://www.unido.org/sites/default/files/2008-07/CSR_-_Implications_for_SMEs_in_Developing_Countries_0.pdf
2. Scholz-Reiter, B., Freitag, M., and Schmieder, A. A dynamical approach for modelling and control of production systems. *AIP Conf. Proc.*, 2002, **622**(1), 199–210.
3. Stricker, N., Micali, M., Dornfeld, D., and Lanza, G. Considering interdependencies of KPIs – possible resource efficiency and effectiveness improvements. *Procedia Manuf.*, 2017, **8**, 300–307.
4. Snatkin, A., Eiskop, T., Karjust, K., and Majak, J. Production monitoring system development and modification. *Proc. Est. Acad. Sci.*, 2015, **64**, 567–580.
5. Paavel, M., Karjust, K., and Majak, J. Development of a product lifecycle management model based on the fuzzy analytic hierarchy process. *Proc. Est. Acad. Sci.*, 2017, **66**(3), 279–286.
6. Anggadwita, G. and Mustafid, Q. Y. Identification of factors influencing the performance of small medium enterprises (SMEs). *Procedia – Social Behav. Sci.*, 2014, **115**, 415–423.
7. Venckeviciute, G. and Subaciene, R. European influence upon Lithuanian SME performance measurement. *Procedia – Social Behav. Sci.*, 2015, **213**, 261–267.
8. Sahno, J., Shevtshenko, E., and Karaulova, T. Framework for continuous improvement of production processes. *Eng. Econ.*, 2015, **26**, 169–180.
9. Lehtimaki, A. Management of the innovation process in small companies in Finland. *IEEE Trans. Eng. Manage.*, 1991, **38** (2), 120–126.
10. Amrina, E. and Vilsi, A. Key performance indicators for sustainable manufacturing evaluation in cement industry. *Procedia CIRP*, 2015, **26**, 19–23.
11. Shahin, A. and Mahbod, M. A. Prioritization of key performance indicators. An integration of analytical hierarchy process and goal setting. *IJPPM*, 2015, **56**, 226–240.
12. S. Kadarsah. Framework of measuring key performance indicators for decision support of higher education institution. *J. Appl. Sci. Res.*, 2007, **3**, 1689–1695.
13. Parmenter, D. *Key Performance Indicators (KPI): Developing, Implementing and Using Winning KPIs*. Second ed. John Wiley & Sons, Inc., New Jersey, 2010.
14. Yuan, J., Wang, C., Skibniewski, M. J., and Li, Q. Developing key performance indicators for public-private partnership projects: questionnaire survey and analysis. *J. Manage. Eng.*, 2012, **28**, 252–264.
15. Podgorski, D. Measuring operational performance of OSH management system. A demonstration of AHP-based selection of leading key performance indicators. *Saf. Sci.*, 2015, **73**, 146–166.
16. Kaganski, S. and Toompalu, S. Development of key performance selection index model. *J. Achiev. Mater. Manuf. Eng.*, 2017, **82**(1), 33–40.
17. Kaganski, S., Paavel, M., and Lavin, J. Selecting key performance indicators with support of enterprise analyze model. In *Proceedings of the 9th International DAAAM Baltic Conference, Tallinn, Estonia, April 24–26, 2014*, 97–102. http://innomet.ttu.ee/daaam14/proceedings/Production%20Engineering%20and%20Management/Kaganski.pdf
18. Eckerson, W. W. Performance Management Strategies: How to Create and Deploy Effective Metrics. ftp://public.dhe.ibm.com/software/data/sw-library/cognos/pdfs/analystreports/ar_peformance_mgmnt_strategies_how_to_create_and_deploy_effective_metrics.pdf
19. Paavel, M., Kaganski, S., Karjust, K., Lemmik, R., and Eiskop, T. Analysis model development to simplify PLM implementation. In *Proceedings of the 10th International Conference of DAAAM Baltic, Industrial Engineering, Tallinn, Estonia, May 12–13, 2015* (Otto, T., ed.). DAAAM Baltic, Tallinn University of Technology, Tallinn, 2015, 69–74.
20. Majak, J., Pohlak, M., Eerme, M., and Velsker, T. Design of car frontal protection system using neural networks and genetic algorithm. *Mechanika*, 2012, **18**(4), 453–460.
21. Karjust, K., Pohlak, M., and Majak, J. Technology route planning of large composite parts. *Int. J. Mater. Form.*, 2010, **3**, 631–634.
22. Lellep, J. and Majak, J. Nonlinear constitutive behavior of orthotropic materials. *Mech. Compos. Mater.*, 2000, **36**(4), 261–266.
23. Aruniit, A., Kers, J., Goljandin, D., Saarna, M., Tall, K., Majak, J., and Herranen, H. Particulate filled composite plastic materials from recycled glass fibre reinforced plastics. *Mater. Sci. (Medžiagotyra)*, 2011, **17**(3), 276–281.
24. Aggarwal, C. C. *Outlier analysis*. IBM T. J. Watson Research Center, Yorktown Heights, NY, 2013.

Ettevõtte analüüsi mudeli optimeerimine võtmenäitajate valikuks

Sergei Kaganski, Martin Eerme ja Ernst Tungel

Käesoleva uurimistöö käigus tutvustati mõõdikute valimi mudelit, mis omakorda võimaldab genereerida vajalikke mõõdikuid ja ettevõtte analüüsi mudeli põhikontseptsioone. Püstitati mudeli optimeerimise probleem, mille eesmärgiks oli leida väikese ja keskmise suurusega ettevõtete kitsaskohad ning mille lahendamiseks töötati välja vastav protseduur. Ettevõtte analüüsi mudeli sorteerimisel kasutati veebitööriista Optimal Workshop. Küsimuste hindamise etapis, kõrvalekallete tuvastamiseks ja eemaldamiseks kasutati järgmisi meetodeid: standardhälve, modifitseeritud Z-skoor, Tukey meetod ja kohandatud boksplotmeetod.

Proceedings of the Estonian Academy of Sciences,
2019, **68**, 4, 376–382
https://doi.org/10.3176/proc.2019.4.05
Available online at www.eap.ee/proceedings

MANUFACTURING ENGINEERING

A simulation study to determine the parameters of medicine inventory policy

Lawrence Moh'd Al-Fandi*, Aishah Abdullah Bin Obaid, Badreyah Ismaeil Alfailakawi, Hoor Ali Alsubaiei, and Sarah Ali Khudhair

College of Engineering and Technology, American University of the Middle East, St. 250, Block 6, Egaila, Kuwait

Received 12 April 2019, accepted 9 May 2019, available online 23 October 2019

Abstract. Inventory management for the healthcare system is a very important issue as it is directly related to human lives. A medical warehouse in Kuwait was facing several challenges related to inventory management. Stock-out or overstock of some medications can occur occasionally. Simulation model using Arena was used to study the current inventory system. The model was modified to propose a new continuous review (s, S) control policy for efficient inventory system to fix monthly supply. Optimization problem was formulated using Arena OptQuest tool in order to determine the optimal parameters for the policy that would give high availability of medicines with the lowest total cost.

Key words: inventory policy, discrete event simulation, optimization, supply chain management, healthcare systems.

1. INTRODUCTION

In healthcare systems, proper inventory management of medicine is very important to avoid stock-out and overstock. Poor inventory systems could increase the cost and/or decrease the availability of stocks. If stock-out occurs, there is no inventory of a certain item to fulfill the customer's order [1] but in case of overstock the number of stock items exceeds customer's demand [2]. In a medical warehouse in Kuwait, medicine inventory system needed improvement as there was often stock-out or overstock of some types of medicines. As there were changes in the demand pattern and the medicine availability from the supplier, modelling and simulation were suitable tools to establish the most appropriate inventory policy for medicine supply chain [3,4]. The objective of this research was to increase the availability of drugs with satisfying demand from the customers and minimizing the overall costs by decreasing the backordering costs, holding costs, and ordering costs.

2. LITERATURE REVIEW

Inventory policies are classified into two major categories depending on the review period. The first category is the continuous review, in which the inventory position is continuously monitored to make

* Corresponding author, Lawrence.Alfandi@aum.edu.kw

a new order. The second category is periodic review – the inventory position is reviewed after fixed period of time. Most applications regarding the healthcare try to optimize the management and financial performance of drugs warehouse [5]. A wide variety of research papers that study the inventory management in supply chain use simulation modelling that consists of three parts: suppliers, central warehouse, and hospitals. The authors of the paper [6] simulated inventory system to identify the optimal reorder point and reorder quantity in order to reduce the inventory level in the warehouse. Other researches constructed a simulation model in order to study the impact of continuous and periodic review policies. The paper [7] evaluated the performance of several inventory policies based on total costs and patients' safety. Krzyzaniak [8] used optimization to address optimal inventory policies for pharmaceuticals in individual care by minimizing the unexpected replenishment orders and maintaining acceptable service level.

3. SYSTEM DESCRIPTION

Current inventory policy uses the continuous review system while demand and lead time are stochastic. At the beginning of each month, the manager checks both, the sales forecast and the available free space in the warehouse. If the quantity of inventory in stock decreases to a certain level, an order is placed to replenish the stock inventory. The order quantity is determined by Eq. (1).

$$\text{Order quantity} = (\text{current average monthly demand} \times 15 \text{ months})\,(1.15) - (\text{stock today} + \text{quantity of upcoming order}), \tag{1}$$

where 1.15 is accounted for 15% of safety stock to avoid having shortages during the lead time. Lead time is the time between placing an order and receiving it to the warehouse. The reorder point is triggered whenever the month in stock drops below fifteen months. Orders of the warehouse according to the Eq. (2) are the following:

$$\text{Months in stock} = \frac{\text{Quantity in stock today} + \text{upcoming orders}}{\text{Monthly average of quantity issued}}. \tag{2}$$

4. DATA ANALYSIS

Considering drugs to be the only stock items, the method was tested on three medicines. Monthly customer demand was collected for 2 years and it was found that each medicine followed a certain pattern and the lead time was either constant or variable. A bar chart was implemented for medicine A as shown in Fig 1. This medicine is used to treat the cold and flu that spreads at a specific time of the year.

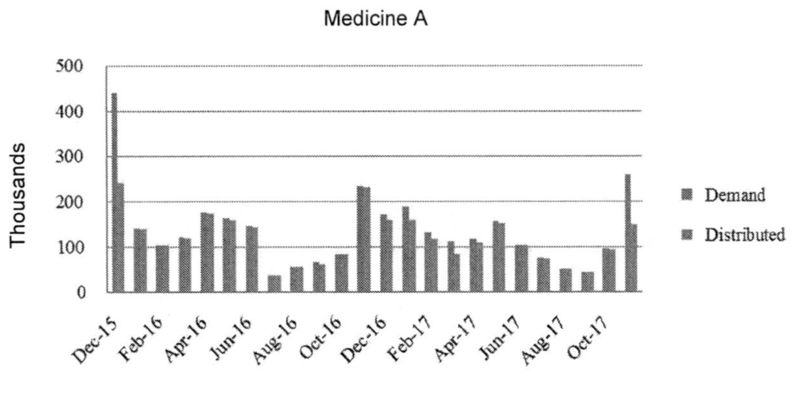

Fig. 1. Medicine A demand behaviour.

It has a low level of criticality as it has many alternatives. There is seasonal demand for that medicine. For example, from August 2016 to October 2016, demand was low as it was not a flu season. From December 2016 to February 2017, demand was high due to the winter season. However, most of the time the warehouse did not meet hospital's demand and that caused a shortage and stock-out in the warehouse.

Medicine B has a moderate criticality – it is important to supply patients with medicine B as it can cause many problems to the patient's health if it is not provided properly. It is used for treating diabetics (adults as well as children). There is need for medicine B all year round. As shown in Fig. 2, a bar chart of medicine B shows that the supplies of the warehouse do not meet the demand of hospitals, which leads to a shortage and stock-out in the warehouse. There is high demand in that medicine almost all year round, which means that medicine B is often needed.

Medicine C has high criticality as its absence during surgery can cause dangerous problems to the patient's health. It is used for decreasing body's immune system while transplanting an organ into the body. As shown in Fig. 3, the bar chart of medicine C illustrates that the warehouse almost met the demand of the hospitals compared to the other medicines. Also, for example in June 2017 and October 2017, demand was shockingly and unexpectedly high since there were many patients having a transplant surgery. Other months like February 2016 and April 2017, demand was low as there were only few patients who needed the medicine.

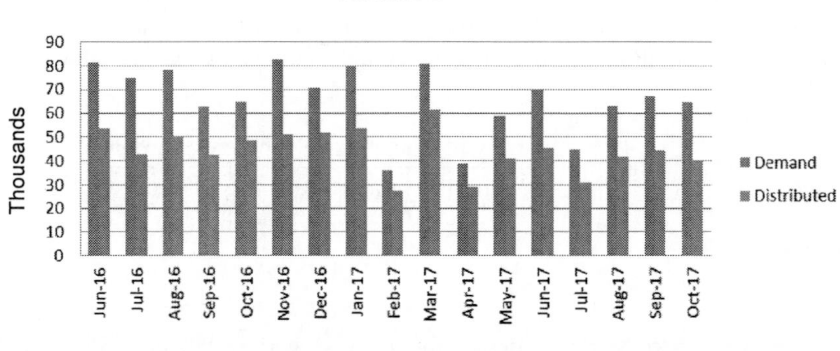

Fig. 2. Medicine B demand behaviour.

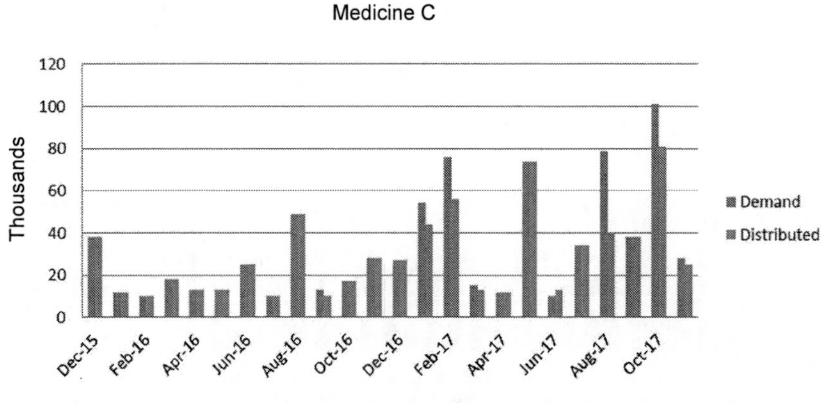

Fig. 3. Medicine C demand behaviour.

5. SIMULATION MODEL

Arena simulation software was used to construct (s_{mos}, S_{mos}) inventory policy, which is (s, S) inventory policy in terms of months of supply (mos) as demand is varying from month to month. (s, S) is a control policy in which the inventory level on hand falls below a minimum (s), replenishment will restore inventory on hand to a target (S). These parameters are specified in number of months. An (s, S) control policy in terms of months of supply identifies that an order will be placed up to (S_{mos}) when the inventory position (IP) falls below the reorder point (s_{mos}). So, when demand (D) occurs, the system checks the availability in stock, if the stock on hand (OH) is enough for the order, the demand will be fulfilled and the quantity on hand is decreased. Otherwise, if the stock on hand is not enough to fill the order, it will be placed as backorder (BO). Backorder will be filled on first-come, first-served basis after the arrival of replenishment order. Inventory position (IP) was calculated using Eq. (3):

$$\text{Inventory position (IP)} = \text{inventory on hand (OH)} + \text{scheduled receipt (SR)} - \text{backorder (BO)}. \quad (3)$$

Every time the order has arrived, the system checks if inventory position (IP) falls below the quantity that is enough for s months of supply (s_{mos}), an order is placed to restore inventory on hand up to S months of supply (S_{mos}). There is certain time from placing an order until it arrives, which is called lead time. After the lead time, the order will be replenished and quantity on hand (OH) is increased. A signal for backorder queue is sent to fill the backorder (BO), and the inventory on hand (OH) is decreased. If the inventory position (IP) is below the reorder point (s_{mos}), a new order will be placed. Figure 4 summarizes the conceptual model for (s, S) policy.

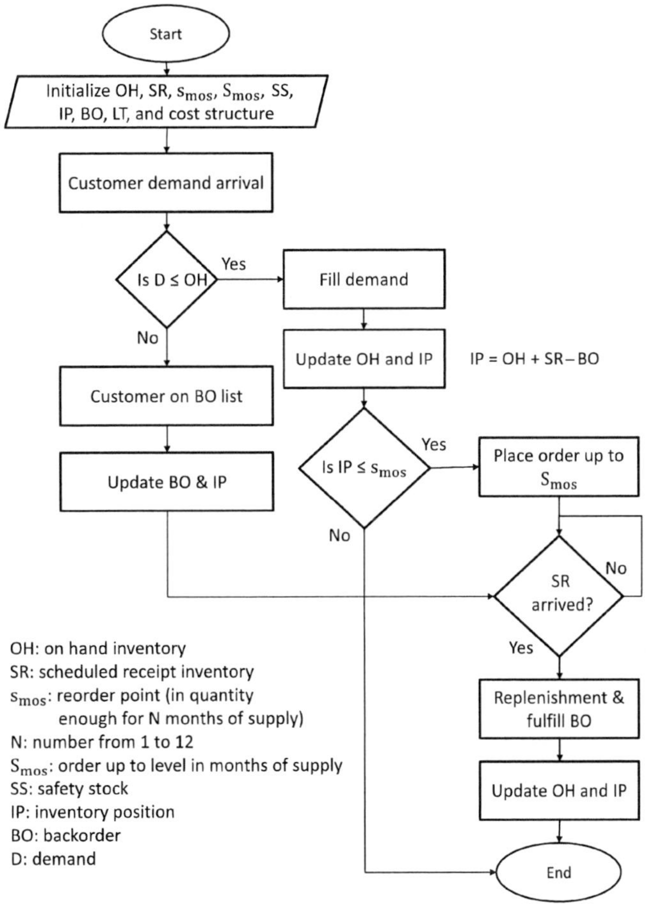

Fig. 4. Conceptual model for (s, S) inventory control policy.

6. PERFORMANCE MEASURES

Two performance measures were identified to represent the performance of the inventory policy: fill rate and total cost. The fill rate is the fraction of customer demand that is met through immediate stock availability. Therefore, the higher the fill rate value, the more customers are satisfied (Eq. (4))

$$\textbf{fill rate} = \frac{\text{demand met}}{\text{demand met} + \text{demand not met}} \times 100, \tag{4}$$

$$\textbf{total cost} = \text{ordering cost} + \text{holding cost} + \text{backorder cost}. \tag{5}$$

Equation (5) shows the total cost, where ordering cost is the cost associated with placing purchase orders and it was estimated to be 70.15 $ per order. Holding cost is the cost associated with storing inventory that remains unsold and it was estimated to be 52% of the unit cost [9]. Backorder cost is the cost associated with not having inventory in stock and it was estimated to be 40% of the unit cost. It increased in case of critical drugs.

7. MODEL VALIDATION

Baseline scenario results of fill rate were compared to the actual results from the real life. Table 1 shows that the process is valid as the actual fill rate for each medicine falls within the confidence interval (CI) generated from the simulation model for the three types of medicines.

8. IMPROVED SCENARIO

After validation of the baseline scenario, the model was modified to enable dealing with different s and S values. The optimal parameters for the control policy that gives the lowest total cost and the highest fill rate were estimated by the Arena OptQuest optimization tool. The goal was to minimize the total cost. There were constraints having (1) the fill rate at least 90% to guarantee sufficient amount of different medicines; (2) $\mathbf{s_{mos}} < \mathbf{S_{mos}}$, as in case the inventory position decreases to or below ($\mathbf{s_{mos}}$) orders will be placed up to the maximum level ($\mathbf{S_{mos}}$); (3) integers for $\mathbf{s_{mos}}$ and $\mathbf{S_{mos}}$ values as orders are placed monthly. The optimal parameters for ($\mathbf{s_{mos}}$, $\mathbf{S_{mos}}$) control policy for each drug were identified as shown in Table 2. From results of optimization, it can be seen that the values of the control policy changed depending on the type of drug. The reason for this variation was different demand pattern of these medicines.

Table 1. Validation test

Type of medicine	Actual fill rate, %	95% confidence interval
Medicine A	62.5	(58.18, 64.18)
Medicine B	91.67	(87.1, 96.1)
Medicine C	75	(73.5, 80.4)

Table 2. Results of OptQuest optimization tool

Inventory policy	Type of medicine		
	Medicine A	Medicine B	Medicine C
s_{mos}	6	4	2
S_{mos}	9	7	5

Table 3. Simulation results

Type of policy	Type of medicine					
	Medicine A		Medicine B		Medicine C	
	Fill rate, %	Total cost, $	Fill rate, %	Total cost, $	Fill rate, %	Total cost, $
Current policy (baseline scenario)	61	12.526	89	90.245	75	2.541
Optimized (s_{mos} , S_{mos}) (improved scenario)	**90***	12.022	92	**87.031***	**93***	2.629

* The difference is statistically significant at $\alpha = 0.05$.

The results of simulation of the total cost and the fill rate for each medicine are shown in Table 3. Simulation modelling was useful for determining the parameters of the inventory system. The results for medicine A and C show significant improvement at $\alpha = 0.05$ in the fill rate with no significant change in the cost. For medicine B, the cost significantly improved with no significant increase in the fill rate.

9. CONCLUSIONS

This research utilized simulation modelling and optimization to determine the inventory policy parameters for different kind of medicines. At first, the simulation model was developed using Arena simulation software to measure the performance of the current management policies. Then, Arena OptQuest optimizer was used to determine the best values of both, s and S that provide high fill rate while minimizing the total cost. The results show statistically significant improvements in the cost and fill rate for each medicine. The approach implemented in this paper could be applied not only in medical fields but also in different fields of industry.

ACKNOWLEDGEMENTS

The publication costs of this article were covered by the Estonian Academy of Sciences and Tallinn University of Technology.

REFERENCES

1. Food Marketing Institute. https://www.fmi.org/ (accessed 2018-12-07).
2. Fernie, J. and Grant, D. B. On-shelf availability: the case of a UK grocery retailer. *Int. J. Logist. Manage.*, 2008, **19**, 293–308.
3. Kelle, P., Woosley, J., and Schneider, H. Pharmaceutical supply chain specifics and inventory solutions for a hospital case. *Oper. Res. Healthcare*, 2012, **1**(2–3), 54–63.
4. Al-Fandi, L., Lam, S., and Ramakrishnan, S. A simulation approach to determine inventory release times for a pull system. In *Proceedings* of *IIE Annual Conference, Cancun, Mexico, June 5–9, 2010* (Johnson, A. and Miller, J., eds), 1–6. https://search.proquest.com/openview/5049e1ee41034c93ea888c1ebe884111/1?pq-origsite=gscholar&cbl=51908 (accessed 2018-10-08).
5. Kammoun, A., Loukil, T., and Hachicha, W. The use of discrete event simulation in hospital supply chain management. In *Proceedings of the 2014 International Conference on Advanced Logistics and Transport (ICALT), Hammamet, Tunisia, May 1–3, 2014.* https://ieeexplore.ieee.org/document/6864108
6. Suraj, S., Sharma, S., and Routroy, S. Positioning of Inventory in Supply Chain Using Simulation Modeling. *IUP J. Supply Chain Manage.*, 2016, **13**(2), 20–32.
7. Gebicki, M., Mooney, E., Chen, S., and Mazur, L. Evaluation of hospital medication inventory policies. *Health Care Manage. Sci.*, 2014, **17**(3), 215–229.
8. Krzyzaniak, S. Model of the impact of parameters controlling replenishment in the BS (min-max) continuous review system on the actual inventory availability. *Sci. J. Logist.*, 2015, **11**(3), 283–294.
9. Rossetti, M. D. *Simulation Modeling and Arena.* John Wiley & Sons, Inc., Hobken, New Jersey, 2016.

Simulatsiooniuuring meditsiinivarude parameetrite määramiseks

Lawrence Moh'd Al-Fandi, Aishah Abdullah Bin Obaid, Badreyah Ismaeil Alfailakawi, Hoor Ali Alsubaiei ja Sarah Ali Khudhair

Tervishoiusüsteemi varude haldamine on väga oluline teema, kuna see on otseselt seotud inimeluga. Kuveidi meditsiiniladu seisis silmitsi mitmete probleemidega, mis olid seotud varude haldamisega. Sageli puudus osal ravimitest laojääk või neid oli laos liiga palju. Praeguse inventeerimissüsteemi uurimiseks kasutati Arena simulatsiooni mudelit. Seejärel muudeti mudelit, et välja pakkuda uus, efektiivne varude haldamise süsteem ühe kuu ulatuses. Optimeerimisprobleem formuleeriti Arena OptQuesti tööriista abil, et määrata selle süsteemi optimaalsed parameetrid, mis annavad kõige väiksemad kogukulud ja ravimite hea kättesaadavuse.

Proceedings of the Estonian Academy of Sciences,
2019, **68**, 4, 383–388
https://doi.org/10.3176/proc.2019.4.06
Available online at www.eap.ee/proceedings

MANUFACTURING ENGINEERING

Suitability analysis of using industrial robots in manufacturing

Tavo Kangru[a*], Jüri Riives[b], Kashif Mahmood[c], and Tauno Otto[c]

[a] Institute of Technology, TTK University of Applied Sciences, Pärnu mnt 62, 10135 Tallinn, Estonia
[b] Innovative Manufacturing Engineering Systems Competence Centre, Teaduspargi 8, 12618 Tallinn, Estonia
[c] Department of Mechanical and Industrial Engineering, Tallinn University of Technology, Ehitajate tee 5, 19086 Tallinn, Estonia

Received 12 April 2019, accepted 9 May 2019, available online 28 October 2019

Abstract. Manufacturing industry robotization is spreading into wider range of processes. Determination if robotization is suitable for the company is one of the most critical issues before selecting industrial robot and designing the robot cell. A survey was carried out among Estonian small and medium sized manufacturing enterprises (SMEs) for this study to determine the utilization of industrial robot (IR) in the industry. More specific study of production unit was conducted, using gathered information, to estimate how the objectives of the production cell design were achieved. The aim of the present scientific work is to map the knowledge whether robotization is suitable or not for the company or working processes and to appoint parameters obtained after using the robot cell for practical manufacturing processes. The study results comprise the suitability assessment method with the set of criteria and key performance indicators (KPIs), that best describe implemented production unit profitability and help SMEs to gain additional economic-technical information for future robot-based unit development.

Key words: industrial robots, feasibility analysis, suitability analysis.

1. INTRODUCTION

Suitability analysis is the process and procedure used to establish a system that meets the needs of users. Suitability of robotization is a basic question for managers who are planning changes in the company. For producing goods, companies have to perform different processes and industrial tasks. There are certain aspects why industrial robots are used for those processes. Theoretically, the basic aspects concern humans, productivity and quality [1–3].

The widest areas using industrial robots are: welding, machine tool servicing, assembling, painting, loading-unloading, packaging, palletizing, and medical applications. Welding ranks among the most important joining processes and has special features for the industrial robots, such as programming task sequences, free definition and parameterization of robot positions/orientations, high repeatability and positioning accuracy of moving paths, high speed of end effector, minimum six degrees of freedom (6DOF), variable payloads depending on the welded products (2–150 kg), advanced programmable logic controllers (PLCs). The vision, strategy and action plans for implementation of robots are described in [4] and statistics about using robots in industry can be found in [5].

Literature review and efficiency analysis of IR cells are important basis for gathering information. To estimate the suitability of using industrial robots in different application areas, it is necessary to analyse the applications of the robot in the industry and to solve a decision-making task.

* Corresponding author, tavo.kangru@tktk.ee

2. BACKGROUND

Suitability analysis belongs to the tasks of dual approach. From one hand, it is an application area for the efficient use of industrial robots and on the other hand it supports decision making methods. The decision maker (expert) must have an excellent understanding about the application area and should be familiar with the factors influencing the effective use of industrial robots. For this purpose, a robot-based manufacturing cell performance evaluation conceptual model was developed, which is based on a recursive decision-making procedure [6]. In the model shown in Fig. 1, there are four groups of parameters: product features, robot cell features, elements of evaluation and general output description. The first two groups are the parameters of the design level (parameters of the product portfolio and their manufacturing processes) and the last two groups are the execution outcomes (different KPIs, that measure, and critical success factors – cost factors, level of achieving the general objectives, dynamics of effectiveness, employee competencies, etc.). These interactions reflect the suitability of using the real IR cell in the company.

The main concern in the suitability analysis process is to find the best solution, according to the set of criteria using the method, which allows the most realistic input (importance) of each criterion. There are different possibilities to influence the roles of criteria: the equal weight (EQW) heuristic [7], the weighted additive (WADD) rule [7]. However, the main risk is over-estimating some of the criteria or not paying enough attention to others. Therefore, artificial intelligence (AI) methods may be used [8].

Analysing industrial robot's applications for welding, there were developed three general groups of parameters, shown in Table 1. Those listed parameters have the greatest influence to the suitability of using welding robots in the company.

Having knowledge about the welding process and parameters make the welding process more efficient. It is possible to find the tools for suitability analysis. To solve the engineering task, such as making a decision about the suitability of welding robots in the company, multiple criteria should be used (see Table 1). Each decision corresponds to a variable, relation or predicate, whose possible values are listed among the condition alternatives. Each action is a procedure or operation

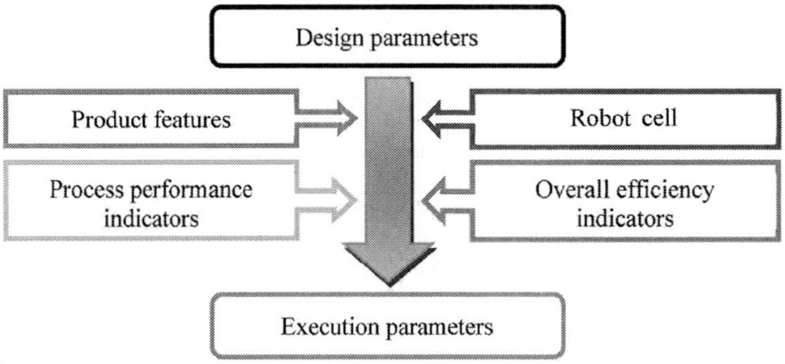

Fig. 1. Efficiency analysis of implementing robot cells in the companies.

Table 1. Suitability criteria for robot welding

Product view	Technology view	Objectives' view
1. The products are complicated from the technological point of view 2. The products can be classified into product families 3. The products are produced in repeatable batches 4. The products are of high quality 5. It is necessary to use welding fixtures	1. Experiences in MIG/MAG and TIG welding 2. Competences in welding technologies 3. Welding processes have great importance in the company's production processes 4. There are already experiences with robot welding 5. It is necessary to increase the productivity of welding processes	1. To shorten the throughput time 2. To increase the productivity in the workplace 3. To increase the product quality 4. To increase the precision of delivery 5. To reduce the product cost

to perform, and the entries specify whether or in what order the actions should be performed for the set of condition alternatives the entry corresponds to.

3. ANALYTIC HIERARCHY PROCESS

Solving the engineering decision making problem of IR cell suitability, which has multiple criteria and alternatives, is a difficult task. One of the techniques for solving multi-criteria decision making (MCDM) problems is analytic hierarchy process (AHP). AHP was proposed by Saaty [9] and developed further by [10,11]. Those methods use fundamental scale of relative importance to construct a pairwise comparison matrix of attributes. Likewise, consistent weight of attributes is determined, which help evaluating composite performance score of alternatives. The alternatives are then ranked according to their composite performance score. Several steps and principles should be considered and understood for constructing a MCDM problem solving tool. The steps are the following:

(1) Developing the hierarchy criteria model for decision;
(2) Deriving priorities by pair-wise comparison for the criteria. Pair-wise comparison scales are shown in Table 2;
(3) Determining local priorities for alternatives;
(4) Calculating and adjusting the consistency:
   (a) Multiplying the matrix of judgements by the eigenvector, obtaining a new vector ($A\omega$);
   (b) Dividing each component of a new vector of $A\omega$ by the corresponding eigenvector element;
   (c) The mean value from the point b is the estimated for λ_{max};
   (d) Calculating the consistency index (CI) by:

$$CI = (\lambda_{\max} - n)/n; \tag{1}$$

   (e) Calculating the consistency ratio (CR) by:

$$CR = CI / RI; \tag{2}$$

   (f) Checking the consistency of the hierarchy. CR should be below or equal to 10 %;
(5) Populating the judgement matrix with input data: quantitative data, such as product payback period total investment etc. are normalized by using Eq. (3). Dimensions, mass etc., are normalized by using Eq. (4). Qualitative data, such as complexity of operations, manufactured parts precision, experience and competencies of engineering stuff and workers, etc., are graded by the scale of 1–5 and normalized.

$$Z = 1 - \frac{x - x_{\min}}{x_{\max} - x_{\min}}, \tag{3}$$

$$Z = \frac{x - x_{\min}}{x_{\max} - x_{\min}}; \tag{4}$$

(6) Making the final decision.

4. CASE STUDY

Production cells were investigated in twenty SMEs with the number of employees ranging from 20 to 150. They produced different parts for agricultural and forestry machines, small tractors, high speed trains, lifts' components, wind generator rotors and other sheet metal products. The information was acquired by interviewing companies' management, engineering staff and data extracted from the enterprise resource planning (ERP) software. Data gained from the interviews and ERP system contained both, quantitative and qualitative data. From the collected data, the information about the robot welding production units was the only one used for the following suitability analysis.

Three performances of production units were stated as a benchmark for the suitability analysis. Production cells, shown in Table 3, were chosen by their excellent KPIs' outcomes. KPIs were selected according to the performance evaluation model [6] and they are as follows: discounted payback period (DPP), cell utilization (CU) and overall equipment effectiveness (OEE).

Table 2. Pair-wise comparison scale assessment

Importance	Description
1	Equal Importance
3	Moderate importance
5	Strong importance
7	Very strong importance
9	Extreme importance
2, 4, 6, 8	Values between two adjacent values should be in considerations
Inverse	If activity (i) got a point compared with activity (j), then (j) has the opposite value compared to (i)

Table 3. Production unit's description and performances

Company	Production cell	Products	Shifts	DPP, years	CU, %	OEE, %
No. 1	Yaskawa IR, two axes positioner	Heat exchangers	2	3	51	72
No. 2	ABB IR, single axes positioner	Trailer frames	2	2	40	70
No. 3	Yaskawa IR, single axes 2-station positioner	Forestry machine frames	1 (2)	3	45	70

5. CRITERIA AND SUB-CRITERIA

Regarding this study, the main task of multi criteria decision analysis (MADM) is to estimate the suitability index and it is based on the following criteria and sub-criteria (see Fig. 2):

- Production unit (PU):
  - cost (C): total investment (C1), cost of utilities (C2), running costs (C3);
  - maintenance (M): maintenance cost (M1), emergency maintenance cost (M2);
  - level (L): use of CAD/CAM (L1), automated storage (L2), machine vision (L3).
- Product (P):
  - physical properties (PP): complexity of parts (T1), parts manufacturing precision (T4), mass (T6);
  - productivity (PR): product families (T2), patch size (T3), patch repeatability (T9), overall welding ratio (TE3), average cycle time (TE9), average setup time (PR2), quality assurance (E2).
- Company environment (CE):
  - workforce (WF): workstation fulfillment (E1), workers salary (E6), production engineer's involvement (E8), shifts (W2), durations of shifts (W3);
  - performance indicators (PI): increment of productivity (E4), increment of on time delivery OTD (E9), increment overall equipment effectiveness OEE (E10), payback period (K1);
  - experiences (E): experiences with MIG/MAG, TIG (TE1), competencies in welding technology (TE2), experiences with robotization (TE4), experiences with jigs and fixtures (TE7), workstation organization level (TE8), overall automatization level (TE10).

6. PERFORMANCE SCORES

The assessments obtained from the decision makers are made by pairwise comparisons. Performance scores and consistency ratio are calculated and given in Tables 4 and 5.

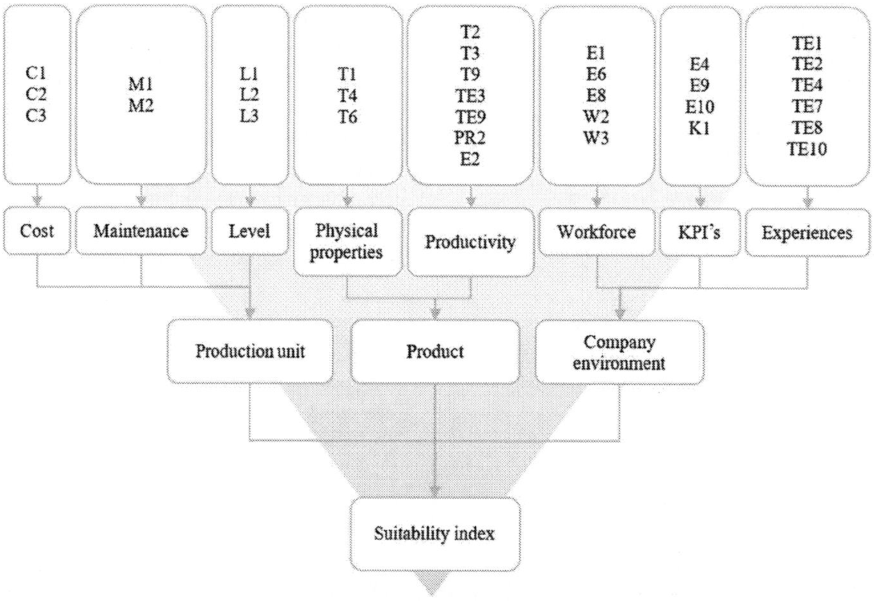

Fig. 2. Production cell suitability hierarchy.

Table 4. Performance scores of main criteria

Criteria	CR	Priority	Criteria	CR	Priority
Production unit		21	Physical properties	0	66.7
Product	1.9	24	Productivity		33.3
Company environment		55	Workforce		19.5
Cost		51.3	Performance	9.8	8.8
Maintenance	5.6	8.1	Experiences		71.7
Level		40.6			

Table 5. Sub-criteria, local performance scores

Criteria	CR	Priority	Criteria	CR	Priority	Criteria	CR	Priority
C1		74.3	T2		4.4	W3	2.6	18
C2	7.4	6.3	T3		18.1	E4		23.8
C3		19.4	T9		35.5	E9	7.6	28
M1	0	66.7	TE3	8.9	9.6	E10		8.9
M2		33.3	TE9		16.6	K1		39.3
L1		26	PR2		7.5	TE1		20.2
L2	5.6	41.3	E2		8.3	TE2		24.5
L3		32.7	E1		25.4	TE4	7.7	20.6
T1	5.6	41.3	E6	2.6	33.9	TE7		8.8
T4		32.7	E8		7.5	TE8		10.1
T6		26	W2		15.2	TE10		15.8

Explanations for the abbreviations are given in paragraph 5. Criteria and sub-criteria.

7. DECISION MATRIX

The normalized inputs are multiplied by their corresponding performance scores and the local and global scores are summed up. Results are shown in Table 6.

8. DISCUSSION AND CONCLUSIONS

In this study an AHP based suitability analysis for robot integrated production cells was developed. Twenty production cells in different industries and at different levels were investigated. Based on the literature, review input parameters were selected, criteria set up and hierarchy of the problem were developed. To ensure the objectivity of experts' pairwise comparisons of the responses of criteria, consistency ratio was calculated and controlled. For testing the developed tool, a case study approach was used. Three welding cells were selected based on their excellent KPIs' outcomes and set as a benchmark for suitability analysis. The highest overall suitability score was obtained in case of No. 3 with index of 0.17. The extremely high score was received in both, product and company environment categories, i.e. 0.849 and 0.810, respectively. The suitability analyses confirmed an excellent choice of product to be produced in a well-organized cell and automated company environment. For decision of suitability, four categories were proposed in Table 7, based on suitability criteria for robot welding, shown in Table 1.

For more precise results, it is possible to simulate the planned robot cell and to calculate the break-even point. Having enough competence in all these areas is quite sophisticated. Therefore, the tool which gives the possibilities to estimate the suitability of using industrial robots for the automation of a certain manufacturing

Table 6. Suitability index results

Production cell	Production unit	Product	Company environment	Suitability index
Company cell No. 1	0.567	0.699	0.664	0.652
Company cell No. 2	0.494	0.709	0.804	0.716
Company cell No. 3	0.524	0.849	0.810	0.717

Table 7. Suitability decision categories

Suitability index	Decision	Description
Smaller than 0.25	No expediency	Products portfolio, analysis of the current process and general conditions are indicating the lack of essential need for using robots in the company.
Smaller than 0.5	To a certain extent expedient	There is indicated the strong point (products, process, general conditions) and also the problematic places. The final decision lays on the industrial expert.
Smaller than 0.75	Robotization is recommended	There are indicated some risks which are not so much important.
Higher than 0.75	Robotizing is feasible	Each group (product, process, manufacturing conditions) has an index higher than 0.75, which gives a solid knowledge that robotization of the process would give significant benefits to the company.

process, is important in the early stage of planning. For future work, more robot integrated processes like machine tending, palletizing, etc. can be added to the tool.

ACKNOWLEDGEMENTS

We thank our colleagues from Taltech, IMECC and TTK UAS who provided insight and expertise that greatly assisted the research. The publication costs of this article were covered by the Estonian Academy of Sciences.

REFERENCES

1. Miller, R. K. *Industrial Robot Handbook*. Springer, New York, 1989.
2. Nof, S. Y. *Handbook of Industrial Robotics, Second Edition*. John Wiley & Sons, New York, 1999.
3. Koren, Y. *Robotics for Engineers*. McGraw-Hill Book Company, New York, 1985.
4. New Robot Strategy. Japan's Robot Strategy. Vision, Strategy, Action Plan. https://www.meti.go.jp/english/press/2015/pdf/0123_01b.pdf
5. World Robotics Report 2016. International Federation of Robotics. https://ifr.org/ifr-press-releases/news/world-robotics-report-2016
6. Kangru, T., Riives, J., Otto, T., Pohlak, M., and Mahmood, K. Intelligent Decision Making Approach for Performance Evaluation of a Robot-Based Manufacturing Cell. In *Proceedings of the International Mechanical Engineering Congress and Exposition, November 9–15, 2018, Pittsburgh, Pennsylvania, USA*. https://asmedigitalcollection.asme.org/IMECE/proceedings-abstract/IMECE2018/52019/V002T02A092/276341
7. Pfeiffer, J. Fundamentals on Decision-Making Behavior. In *Interactive Decision Aids in E-Commerce*. Physica-Verlag HD, Berlin, Heidelberg, 2011, 15–45. https://link.springer.com/chapter/10.1007%2F978-3-7908-2769-9_2
8. Zhang, L., et al. Using Neural Network to Evaluate Construction Land Use Suitability. In *Proceedings of the 2010 Second International Workshop on Education Technology and Computer Science, March 6–7, 2010, Wuhan, China*. https://ieeexplore.ieee.org/document/5459823
9. Saaty, T. L. *The Analytic Hierarchy Process*. McGraw-Hill, New York, 1980.
10. Kaganski, S., Majak, J., and Karjust, K. Fuzzy AHP as a tool for prioritization of key performance indicators. *Procedia CIRP*, 2018, **72**, 1227–1232.
11. Paavel, M., Karjust, K., and Majak, J. Development of a product lifecycle management model based on the fuzzy analytic hierarchy process. *Proc. Est. Acad. Sci.*, 2017, **66** (3), 279–286.

Tööstusrobotite kasutatavuse sobivusanalüüs

Tavo Kangru, Jüri Riives, Kashif Mahmood ja Tauno Otto

Tootmisettevõttes on määrava tähtsusega enne robottootmisüksuse loomist läbi viia tööstusrobotite sobivusanalüüs. Selle väljatöötamiseks tehti Eesti väikese ja keskmise suurusega ettevõtete hulgas uuring, määramaks tööstusrobotite kasutust. Kogutud andmete põhjal viidi läbi spetsiifilisem tootmisüksuste uuring, millega hinnati tootmisrakkude projekteerimisel püstitatud eesmärkide saavutamist. Tulemusena loodi tööstusrobotite sobivuse hindamise meetod koos kriteeriumide ja tulemuslikkuse võtmenäitajate kogumiga. Hindamismeetod võimaldab hinnata rakendatud tootmisüksuse kasumlikkust ja saada täiendavat majanduslik-tehnilist teavet tulevaste robottootmisüksuste arendamiseks.

Proceedings of the Estonian Academy of Sciences,
2019, **68**, 4, 389–394
https://doi.org/10.3176/proc.2019.4.07
Available online at www.eap.ee/proceedings

MANUFACTURING ENGINEERING

Integration of autonomous vehicles and Industry 4.0

Raivo Sell[a*], Anton Rassõlkin[b], Ruxin Wang[a], and Tauno Otto[a]

[a] Department of Mechanical and Industrial Engineering, Tallinn University of Technology, Ehitajate tee 5, 19086 Tallinn, Estonia
[b] Department of Electrical Power Engineering and Mechatronics, Tallinn University of Technology, Ehitajate tee 5, 19086 Tallinn, Estonia

Received 12 April 2019, accepted 16 May 2019, available online 12 November 2019

Abstract. The current trend of automation and manufacturing is digitalization called Industry 4.0. The digitalization is not only the manufacturing process upgrade but also logistics and intelligent transport solutions. Rapid development of intelligent control techniques has also brought changes in automotive industry and led to developments of autonomous vehicles. This paper presents a concept for an integration of self-driving vehicle ISEAUTO into Industry 4.0. ISEAUTO is a last mile automated shuttle bus designed and built in Estonia for short range transportation.

Key words: self-driving car, Industry 4.0, autonomous shuttle.

1. INTRODUCTION

Autonomous vehicles (AVs) and Mobility-as-a-Service (MaaS) can provide more efficient logistics and passenger transfer methods. Furthermore, smart sensors along with implementation of cognitive computing and the Internet of Things (IoT) represent an AV as a cyber-physical system (CPS), within which information from all related perspectives is closely monitored and synchronized between the physical devices and the cyber computational space. By utilizing advanced information analytics, AVs will be able to perform more efficiently, collaboratively and resiliently. Thus, it is possible to integrate AVs into Industry 4.0 systems. Competence and know-how about AVs as well as self-driving algorithms gained during ISEAUTO project presents a base for urban mobility studies. The paper presents ISEAUTO as a CPS that can be abstracted into sensing, computing, and actuation modules.

According to the report from McKinsey & Company, fully automated vehicles are unlikely to be commercially available before 2020 [1]. The implementation of AVs combing the application of MaaS or Transport-as-a-Service (TaaS) have become more realistic and meaningful. The readiness of these technologies empowers researchers to help industries to increase their productivity and create more benefits.

Since the term Industry 4.0 was first pushed by German government [2], it has been applied to a larger scope of technical elements and brought more connections between the virtual world and the physical one. Industry 4.0 describes a set of concepts that can drive the next industrial revolution. The potential of adopting the AVs into industrial area is promising for accommodating most technical concerns in the following areas:

- *Cost-efficient*: instead of traditional fuels, energy like electricity can reduce the cost and increase the effectiveness of manufacturing. Unmanned methods can also help to cut down expenses of employers' payments;

* Corresponding author, raivo.sell@taltech.ee

- *Digitalization*: using cloud computing to deal with big data captured by sensors from AVs help factories to keep control of all information and stay on track of the current status;
- *Safety/Liability*: real-time tracking and monitoring system ensure the safety during the vehicle operation. The avoidance of human intervention ensures better technology acceptance and a sustainable customer relationship;
- *Flexibility/Scalability*: involving more partners in this ecosystem brings more robustness and lowers the risk of dependency on a certain component, reaching continuously to different levels of investments and making the market more competitive;
- *Environment-friendly*: the reduction of the emissions of waste/pollution protects the living environment and adds satisfaction to surrounding residences. Clean energy also contributes to the whole production life circle thus completes the supply chain.

2. INDUSTRY 4.0 IN TERMS OF SELF-DRIVING VEHICLES

Self-driving vehicles are also foreseen to become a part of Industry 4.0. European technology platform ManuFuture[1] is developing a vision for 2030, where production logistics is integrated through self-driving mobile platforms both indoors and outdoors by artificial intelligence (AI), see Fig. 1.

Several initiatives are started in the world to integrate automated shuttles into industrial environment. However, most of the trials are still in early stages. Following use cases are reviewed in more details, a summary is presented in Table 1.

Foxconn[2]. In 2017, Taiwanese firm Foxconn planned to build a $10 billion plant to the south of Milwaukee. They proposed that some lanes should be reserved only for AVs and the driverless-vehicle lanes could connect the Foxconn factory with Milwaukee's Mitchell

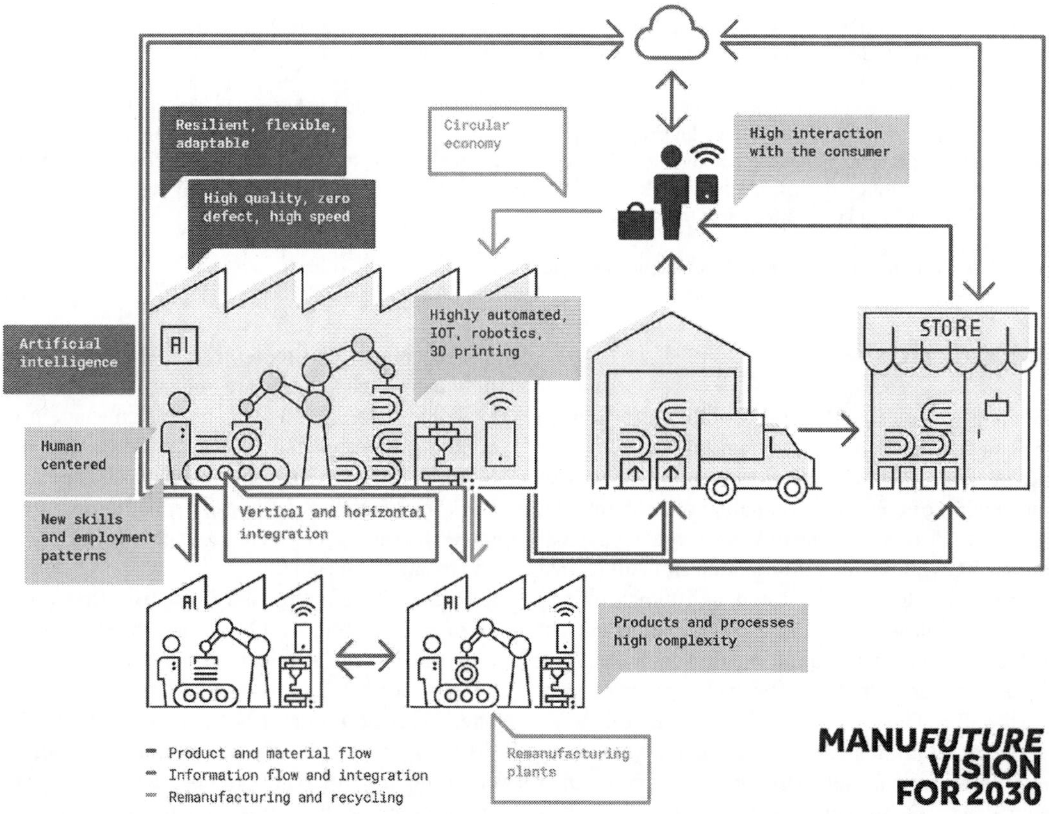

Fig. 1. ManuFuture technology platform vision for 2030.

[1] http://www.manufuture.org/
[2] http://www.foxconn.com/

Table 1. Comparison of shuttle buses developed worldwide

Case study	Location	State-of-art features	Status
Foxconn Factory	Milwaukee, US	Reserved lanes on normal highways only for autonomous vehicles.	Plan
Baidu	Fukushima, Japan	Transport workers within the nuclear power plant facility.	Plan
Tsintel Technology	Tianjin, China	Provide commuting services at a low speed in semi-open environments.	Pilot
Magride	Guangdong, China	Serve as a short-range transportation transferring employees.	Operate

International Airport. This plan would help Foxconn to automate its shipping of both goods and people. While accidents in US involving AVs have attracted attention in headlines, the advantage of such vehicles is safety, especially on dedicated routes with a constant ebb and flow of users.

Baidu[3]. Under the agreement with SB Drive, ten Baidu Apolong mini buses will be exported to Japan from China in early 2019. Apolong buses are set to enter Japan's self-driving market as shuttle buses in practical use at the crippled Fukushima No. 1 nuclear power plant to transport workers within the facility. There is also a plan of using self-driving bus service for maintenance workers and tourists at Tokyo Haneda Airport.

Magride[4]. Magride is a leading solution provider for short-range intelligent campus transportation in closed areas like airports, industrial plants, residential quarters and amusement parks. Its L4 shuttle bus G100 has already been used in Vanke Architecture Research Centerinstitute (Guangdong, China) for more than 1 year as a short-range transportation transferring employees. Its latest product G200 started pilot tests on open roads since August 2018.

Tsintel Technology[5]. Their pilot tests started from August 2017 in Tianjin Huaming High-Tech Industrial Park (CHINA) with an L4 autonomous shuttle bus, operating in semi-open environments. The shuttle bus can provide commuting services at a low speed on fixed and controllable routes. It can also be managed as a fleet for campus commuting, logistics, road cleaning or even excursion tours.

3. ISEAUTO AS A LAST MILE PLATFORM FOR INDUSTRY 4.0

Nowadays the most effective autonomous approach relies on data fusion of information, coming from plenty of different sensors, some of these sensors are already available on commercial cars [3], others will be introduced step-by-step. The research and development activities have to enable high/full autonomous driving; and Industry 4.0 has to deal with a lot of new requirements (e.g., fail-operational systems, cyber security), technologies (connectivity over 5G, neural networks, future computing platforms), and topics (data analytics, AI) [4].

ISEAUTO [5] last mile bus is operating in the campus of Tallinn University of Technology (TalTech) close to parking lots and some pedestrian areas. That means that an obstacle should be detected in both short and long distances at the same time. It is difficult to have an accurate and fast prediction of the pedestrian behaviour in case of the car in order to react properly. In case of a stop or a recalibration of sensors, situations can occur when a pedestrian is not detected properly due to overlapping with another pedestrian. In case of a jaywalker the system must be able to react and remain in a safe state. This holds also for people lying on the ground or small animals. In all possible situations the system has to be able to stop before any injury occurs. Sensors shown in Table 2 are used in ISEAUTO last mile bus to observe the environment and for mapping, localization and navigation processes.

Similar sensors are often used in industrial systems. For example, machine vision is widely implemented in adaptive industrial robots [6] and LiDARs for safety curtains.

With many national/regional schemes implemented, the autonomous vehicle market is getting more and more crowded these years. Year 2017 and 2018 have witnessed many players increasing their involvement in this global competition; and huge investments from all over the world chasing the same goal and trying to get a share from the market. A successful application of using AV as a service in industry needs to take certain criteria into consideration (see Fig. 2).

[3] http://ir.baidu.com/
[4] http://www.magride.cn/
[5] http://www.tsintel.net/

Table 2. Sensors used in ISEAUTO last mile shuttle bus

Sensor	Type	Quantity	Placement
LiDAR	Velodyne VLP-16	2	Roof
LiDAR	15° VFoD	1	Front
Ultrasonic	PGA450-Q1	8	4 x front, 4 x back
Radar	IWR1443	1	Front
Camera	Basler	5	2 x front, 2 x sides, back
Satellite navigation	RTK-GNSS+IMU	1	Roof

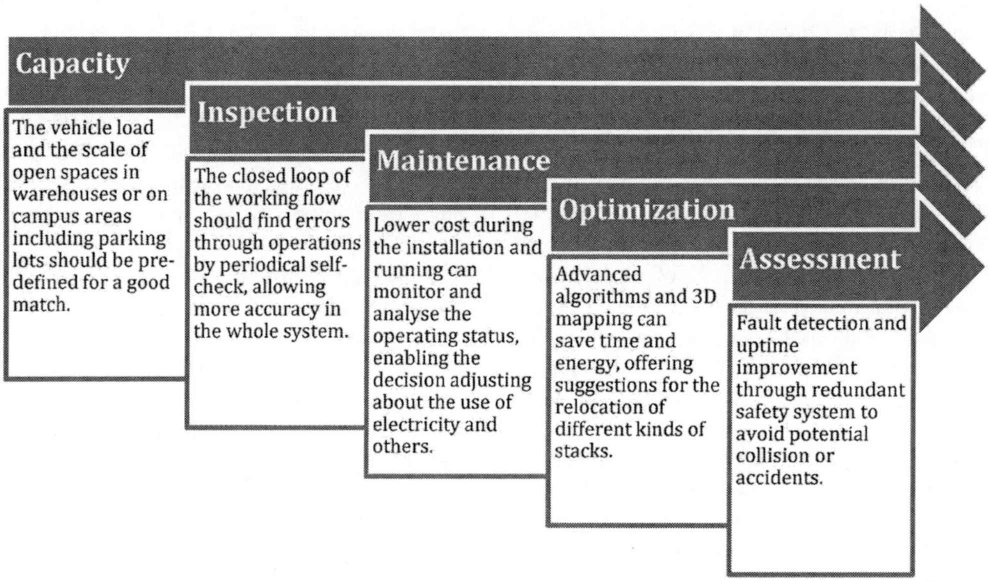

Fig. 2. Criteria for using autonomous vehicles as a service for industry.

The ISEAUTO navigation technology doesn't require much fixed installation of infrastructure. It can be tailored to specific requirements in a cost-effective method, even for industry applications, from individual steps in a process to complex material flows. The advanced technology of the self-driving platform creates significant competitive advantages, while guaranteeing a high flexibility in operational procedures at the same time. Safety systems and an integrated system are a prerequisite for a safe and efficient human-machine interaction (HMI).

The unmanned operation is feasible in many industrial fields, for either individual or defined fleet operations. The automated vehicles are not tied to shift patterns, thus substantially increasing cost-effectiveness, while bringing reductions in daily operational damage. Despite of highly autonomous functions, the whole range of models can also be easily operated using conventional manual controlling mode and can well cope with potential technical issues like power cut-off, network failure etc.

Autonomous multi-agent systems, which optimize the planning and scheduling of industrial processes using the example of courier and express services is presented in [7], where the multi-agent-based approach is used for the optimization and synchronization of logistic and production processes in Industry 4.0 applications. Another research approach that nowadays receives attention in the industry is cellular transport systems. The idea of cellular transport systems [8] is embodied by dedicated (cellular) material handling entities. Cellular transport systems can be represented by autonomous conveying modules or transport vehicles. They are built upon a controlling architecture consisting of several small, self-organizing, intelligent units, which act and decide autonomously [8].

4. AUTOMATED VEHICLES IN INDUSTRY 4.0

The proposed concept is integrating autonomous shuttle vehicles with multi-agent based workflow. The AI supported smart planning and scheduling of production involves self-driving vehicles into the process to serve intelligent resource transportation. In industry, performance indicators and energy efficiency are the most important factors to automate production [9]. These factors can be improved by smart production planning and scheduling service, which has all information from the production. It is important that all the equipment is connected to the cloud and provides real-time information about energy consumption, productivity and capacity. In case of the lack of this functionality, a monitoring system can be added also for older equipment as discussed in detail in [10]. By extending the production planning and scheduling with self-driving vehicle, much higher flexibility and efficiency can be achieved. Figure 3 is presenting the integration concept where production cells are physical units located all over the industrial area of the factory. One or more autonomous vehicles (AV 1, AV n) are serving the production units based on resource requests and guided commands from a cloud-based planning and scheduling system.

In the concept, several types of AVs can be applied for different purposes. Last mile shuttle buses like ISEAUTO are for transporting humans; universal mobile robot like [11] is for transporting goods and materials. The latter is also applied for different purposes in addition to transportation. For example, road cleaning or snow blowing during the idle time or securing the factory area in the night time.

5. DISCUSSION/CONCLUSIONS

In the near future, self-driving vehicles will take its place in everyday life and industry. Requirements from the Industry 4.0 are specified and the very first attempts of integration self-driving vehicles are reviewed in the paper.

ISEAUTO is the first research and educational project in Estonia targeted on the design and development of a self-driving vehicle. The potential of adopting ISEAUTO self-driving platform into Industry 4.0 is proposed in the paper. The long-term objective of the ISEAUTO project is to establish a smart city testbed where different types of rescarches regarding future urban mobility can be studied. The smart city testbed is a real-life environment where self-driving cars, delivery robots and smart infrastructure objects are placed. ISEAUTO last mile bus can be a base for Vehicle-to-everything (V2X) platform, a vehicular communication system that incorporates other more specific types of communication as Vehicle-to-infrastructure (V2I), Vehicle-to-vehicle (V2V) and Vehicle-to-pedestrian (V2P); V2D (Vehicle-to-device), V2G (Vehicle-to-grid), or any other entity that may affect the vehicle.

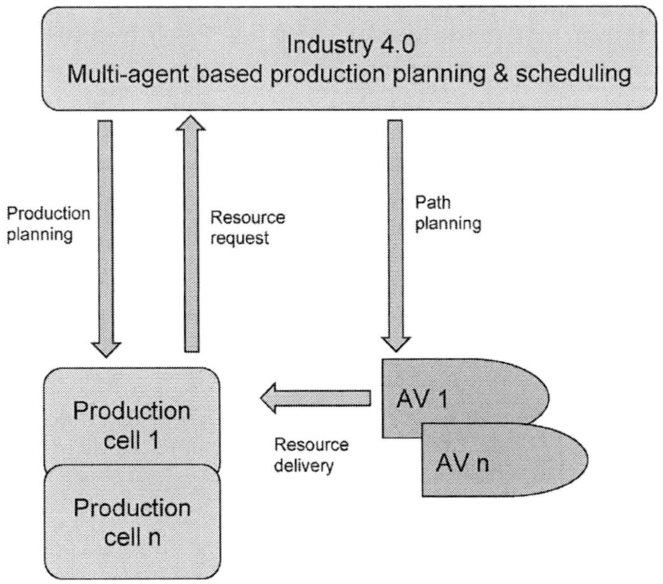

Fig. 3. Integration and production planning and scheduling with autonomous vehicles.

ACKNOWLEDGEMENTS

The project is supported by the TalTech and SilberAuto joint grant SS419. The publication costs of this article were covered by the Estonian Academy of Sciences.

REFERENCES

1. Gao, P., Kaas, H. W., Mohr, D., and Wee, D. Disruptive trends that will transform the auto industry. McKinsey & Company. https://www.mckinsey.com/industries/automotive-and-assembly/our-insights/disruptive-trends-that-will-transform-the-auto-industry
2. Kagermann, H., Lukas, W.-D., and Wahlster, W. Industrie 4.0: Mit dem Internet der Dinge auf dem Weg zur 4. industriellen Revolution. *VDI nachrichten*, 2011, **13**(11), 2.
3. Russo, G., Baccaglini, E., Boulard, L., Brevi, D., and Scopigno, R. Video processing for V2V communications: A case study with traffic lights and plate recognition. In *Proceedings of 2015 IEEE 1st International Forum on Research and Technologies for Society and Industry Leveraging a better tomorrow (RTSI), Turin, Italy, September 16–18, 2015*. IEEE, 2015, 144–148. https://ieeexplore.ieee.org/document/7325064
4. Traub, M., Vögel, H.-J., Sax, E., Streichert, T., and Härri, J. Digitalization in automotive and industrial systems. In *Proceedings of the 2018 Design, Automation and Test in Europe Conference and Exhibition (DATE), Dresden, Germany, March 19–23, 2018*. IEEE, 2018, 1203–1204. https://ieeexplore.ieee.org/document/8342198
5. Rassõlkin, A., Sell, R., and Leier. M. Development Case Study of the First Estonian Self-Driving Car, ISEAUTO. *Electr. Control Commun. Eng.*, 2018, **14**(1), 81–88.
6. Kuts, V., Otto, T., Tähemaa, T., Bukhari, K., and Pataraia, T. Adaptive Industrial Robots Using Machine Vision. In *Proceedings of the ASME 2018 International Mechanical Engineering Congress and Exposition (IMECE2018), Pittsburgh, PA, USA, November 9–15, 2018*. ASME, 2019. https://doi.org/10.1115/IMECE2018-86720
7. Gath, M., Herzog, O., and Edelkamp, S. Autonomous and flexible multiagent systems enhance transport logistics. In *Proceedings of the 2014 11th International Conference and Expo on Emerging Technologies for a Smarter World (CEWIT), Melville, NY, USA, October 29–30 2014*. IEEE 2014, 0–5. https://ieeexplore.ieee.org/document/7021143
8. Schieweck, S., Kern-Isberner, G., and Ten Hompel, M. Using answer set programming in an order-picking system with cellular transport vehicles. In *Proceedings of the 2016 IEEE International Conference on Industrial Engineering and Engineering Management (IEEM), Bali, Indonesia, December 4–7, 2016*. IEEE 2016, 1600–1604. https://ieeexplore.ieee.org/document/7798147
9. Kaganski, S., Majak, J., Karjust, K., and Toompalu, S. (2017). Implementation of Key Performance Indicators Selection Model as Part of the Enterprise Analysis Model. *Procedia CIRP*, 2017, **63**, 283–288. https://doi.org/10.1016/j.procir.2017.03.143
10. Eiskop, T., Snatkin, A., and Karjust, K. Production monitoring system with predictive functionality. *J. Eng. Sci. Technol.*, 2017, **12** (9), 2410–2425.
11. Sell, R. and Otto, T. Remotely controlled multi robot environment. In *Proceedings of the 19th EAEEIE Annual Conference, Tallinn, Estonia, June 29–July 2, 2008*. IEEE 2008, 20–25. https://ieeexplore.ieee.org/document/4610152

Isejuhtivate sõidukite ja Tööstus 4.0 integratsioon

Raivo Sell, Anton Rassõlkin, Ruxin Wang ja Tauno Otto

Tööstuse automatiseerimise ja digitaliseerimise järgmist sammu nimetatakse tööstusrevolutsiooniks Tööstus 4.0. Digitaliseerimine ei tähenda selles kontekstis mitte ainult tootmisprotsesside täiustamist, vaid ka tootmislogistika ja transpordi uusi integreeritud lahendusi. Autotööstuse ja eriti isejuhtivate sõidukite ning autonoomsete süsteemide kiire areng võimaldab luua täiesti uusi Tööstus 4.0 lahendusi. Artiklis on uuritud erinevaid lahendusi, mida maailmas on püütud rakendada, ja välja pakutud TalTechi ning SilberAuto koostöös valminud isejuhtiva platformi ISEAUTO kontseptsioon Tööstus 4.0 integreerimiseks, samuti analüüsitud selle tootmislogistikasse rakendamise võimalusi.

Proceedings of the Estonian Academy of Sciences,
2019, **68**, 4, 395–400
https://doi.org/10.3176/proc.2019.4.08
Available online at www.eap.ee/proceedings

MANUFACTURING
ENGINEERING

Modular smart control system architecture for the mobile robot platform

Raivo Sell*, Eero Väljaots, Tengiz Pataraia, and Ehsan Malayjerdi

Department of Mechanical and Industrial Engineering, Tallinn University of Technology, Ehitajate tee 5, 19086 Tallinn, Estonia

Received 12 April 2019, accepted 16 May 2019, available online 19 November 2019

Abstract. The paper is focusing on the research of current technologies of multi-purpose mobile robots, including implementation of modular layout, open software frameworks, self-organization with plug and play (PnP) capability and adaptive sensor fusion. The hardware layout is studied in detail and modular architecture is proposed as an optimal solution of module interconnection. Software modular architecture is developed in the similar principle, taking plug and play connectivity into account. The framework of the software consists of implementing middleware for the software modularization and three-level hierarchical structure. Practical implementation platforms are introduced and future developments discussed.

Key words: mobile robot, unmanned ground vehicle, system architecture.

1. INTRODUCTION

Unmanned ground vehicles (UGVs) used to be operated by the remote control in performing hazardous tasks over the distance. Nowadays, UGVs, are developed for various purposes, e.g., for research and industrial use, and these are sophisticated vehicles having semi-autonomous or full autonomous modes. For autonomous navigation and safe driving more sophisticated software and hardware is required, including high performance sensors like light detection and ranging (LiDAR); and cameras. Autonomous driving requires different sub-tasks to be solved, e.g. localization, base-navigation, local and global planning, obstacle avoidance, etc. [1]. Although these tasks are similar for most of the mobile robots, they still are very dependent on the selected sensors and robot hardware specifics. To overcome this issue, modular structure of the system is the most obvious choice, especially if the similar functionality has to be applied for different mobile robots. The modular concept of the mobile robots has been studied in many researches and specific solutions have been offered, e.g. in [2,3]. This research is based on the knowledge and expertise acquired from the first Estonian self-driving car – ISEAUTO [4,5], which was put on the road after less than a year-long development. The success of the project motivates to apply similar software and hardware concept also to smaller off-road vehicles. The conceptual solution is an open source modular smart control system for mid-size off-road mobile robotic platforms. Two different robots were used for the experiments and implementation of the concept. The manuscript is organized so that three following chapters introduce two mobile robots as a base platform for the concept implementation. Chapters 5 to 7 describe the concept of the modular architecture and the implementation.

2. IMPLEMENTATION PLATFORMS

The unmanned robotic platforms where the concept of modular architecture is applied, are universal mobile

* Corresponding author, raivo.sell@taltech.ee

robotic platform UKU – developed by the students of Tallinn University of Technology (TalTech); and Agronaut – a robot for the agriculture, developed by the Estonian company Hecada. Both vehicles are in the same size but use different concepts of steering and power system.

3. UNIVERSAL MOBILE ROBOT UKU

The all-terrain mobile robot UKU, shown in Fig. 1, is powered by the electric motor and Li-Io batteries. The robot is with a rear-wheel drive without differential and mechanical transmission. It has off road tires suitable for climbing over the obstacles or for example, plowing snow autonomously on the parking lot in winter. The robot has a special self-contained measurement system to measure the efficiency, similar to [6] and dynamics of energy consumption. Robot weights 260 kg and has nominal speed of 4 m/s. The main electric motor is a permanent magnet DC motor producing 4 kW. Basic navigation sensors are SICK 2D laser scanners (measuring range up to 80 m), ultrasonic rangers on both sides and long-range 360-degree LiDAR on the top of the vehicle. The more detailed architecture of the robot is described in [7].

4. AGRICULTURAL MOBILE ROBOT AGRONAUT

The Agronaut [8], shown in Fig. 2, is a universal mobile robotic platform, which purpose is practical testing of unmanned technologies and navigation in agricultural conditions. Its physical layout is symmetric and modular, consisting of identical modules that are connected to each other through hydraulic steering linkage. One module accommodates the 15-kW power plant that powers the hydraulic system. The other module is free for transporting necessary task-specific equipment. It is also possible to connect the third and the fourth module and actuators using the same physical interface on both

Fig. 1. Universal mobile robot UKU.

Fig. 2. Agronaut UGV platform, modular electronic units on top of the hydraulic system.

ends of the body control module. Agronaut is all-wheel drive robot and has robust design to suit for the agricultural field of use. Therefore, unlike robot UKU, all actuators outside the body are hydraulically powered and hydraulic lines are routed from one unit to another. As the vehicle can be assembled with task-specific modules, the control system also has to be modular and easily configurable for the required task. Each hydraulic actuator has its own electronic control, implementing PID-regulators for controlling hardware, connected with central computer through controller area network (CAN). As the platform weighs 470 kg, it suits perfectly for automating repetitive simple tasks usually carried out by humans, e.g., automated soil sampling of the cultivated land.

5. CONCEPT OF MODULAR ARCHITECTURE

Hardware and software architecture were developed based on technical requirements which were set initially according to the level of functionality and robustness of the system. Key priority of software and hardware architecture was to create a safe, easily reconfigurable and scalable system where each module/microcontroller has a task to control each locomotion unit (steering, driving motors, brakes) and different sensors across the platform. The whole control system is based on two communication protocols – Controller Area Network (CAN) bus and Universal Datagram Protocol (UDP). These protocols were chosen based on several key priorities: speed, reliability, and robustness. Intermodule communication was developed based on the CAN bus. Choice of CAN bus was made due to its high-transmission reliability, real-time capabilities, and robustness. Communication between the master controller and the computer was done through UDP messages due to its speed, reliability, and efficiency. Figure 3 shows SysML block definition diagram (bdd) and internal block diagrams (ibd) for general modular architecture. These diagrams describe how messages are delivered from the main computing unit to the lower level controllers and main parameter values of the data flow.

6. IMPLEMENTATIONS OF MODULAR ARCHITECTURE

The concept of modular architecture is applied to the base platforms described earlier. The hardware architecture implemented in mobile robot UKU has two-level software control, and three-level hardware control architecture. Figure 4 shows implementation of the hardware architecture in specific UGV platform. It has high-level sensors dealing with navigation and obstacle avoidance. High-level sensors are directly connected to the main computing unit, which runs the Robotic Operating System (ROS) middleware and open source components for its main tasks. ROS-based system control algorithms produce driving commands to low-level controllers through the master controller. Master controllers prioritize and translate messages between high-level and low-level controllers. Low-level controllers are dealing with direct control of motor drivers, consisting of proportional-integral-derivative (PID) and ground safety algorithms. There is a separate safety controller and it is independent from both high-level and low-level controllers. It monitors output signals as well as CAN network messages and can stop the vehicle in case anomalies occur. The system has also remote link over the 4G/5G mobile network to provide online data stream between server and operator.

Control software for each control unit of the mobile robot was designed in a similar manner to make it simple, easily understood and configurable in the future for different systems, e.g. front and rear locomotion units of the robot have the same PID regulators with different data inputs and controller coefficients.

Let's take an example of a steering controller where desired wheel angle (setpoint) is sent from the ROS

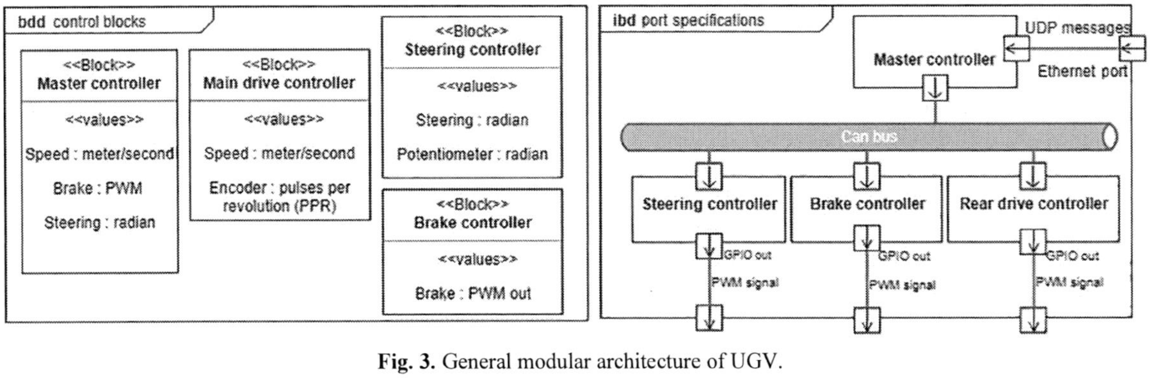

Fig. 3. General modular architecture of UGV.

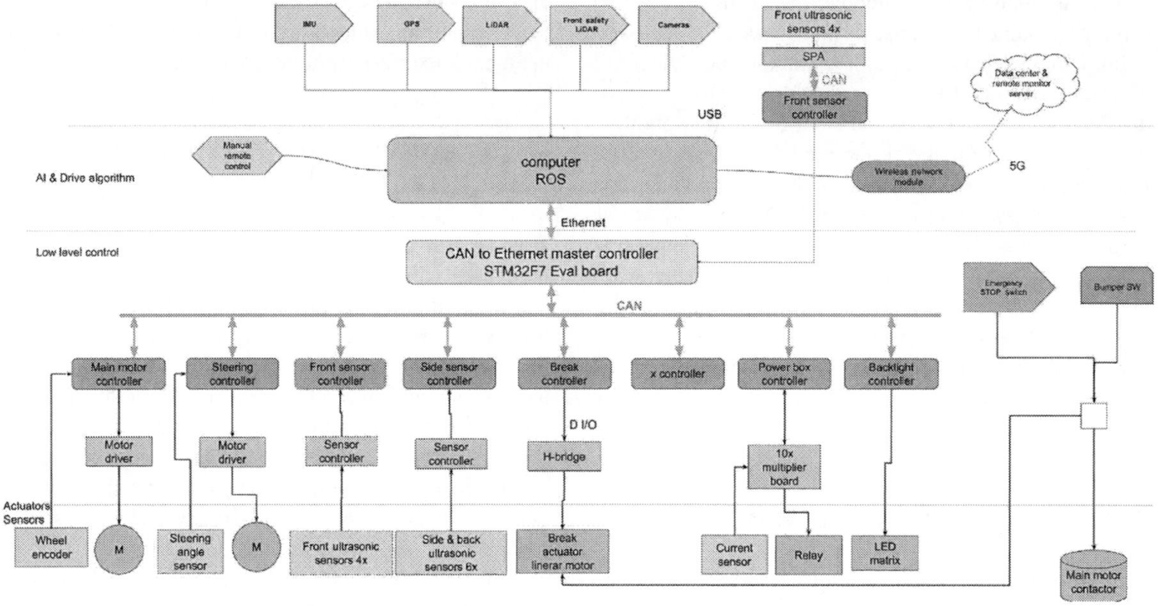

Fig. 4. Hardware architecture implemented on the UGV UKU.

computer with the CAN bus protocol and the controller's task is to calculate the most optimal motor speed to reach the desired steering angle.

Control object, in this case, is H-bridge type motor driver, which is changing the polarity of the voltage based on the pulse width modulation (PWM) signals, which are generated from the controller. Feedback device for the steering motor is a simple analog sensor (potentiometer) due to its simplicity of integration and accuracy. Control process of the PID is shown in Fig. 5.

PID controller of both actuators was tuned and validated separately by step tests. This process clearly shows the dynamical characteristic of each actuator. Testing and fine-tuning of PID controller defines the overall performance of the mobile robot in the future and is therefore crucial. Figure 6 shows the most optimal test result for front steering. Setpoint of the experiment was steering angle input from the computer varying from –1.0 to 1.0 radians, and potentiometer value as a controller feedback. This test case where the practical method of [9] was followed shows step response for controller with an angle input from 0.9 rad to –0.9 rad, which is one full rotation of the steering axle of the UGV. Proportional derivative (PD) action of the PID controller was enough to reach the optimal results for both steering and rear-wheel drive. The advantage of using only PD characteristics is rapid output and short time required to return process value to setpoint. PD formula can be seen in Eq. (1):

$$u = k_p + e + k_d \frac{d_e}{d_t}, \qquad (1)$$

where u – static characteristics, k_p – proportional gain, k_d – derivative gain, e – error, d_e – change in error; and d_t – change in time. After tuning the PD parameters for steering motor we got the most optimal result using coefficient values: $k_p = 0.1$ and $k_d = 0.9$. The graph shows that the steady state is reached after 8 seconds. This means that the robot will do one full turn of the

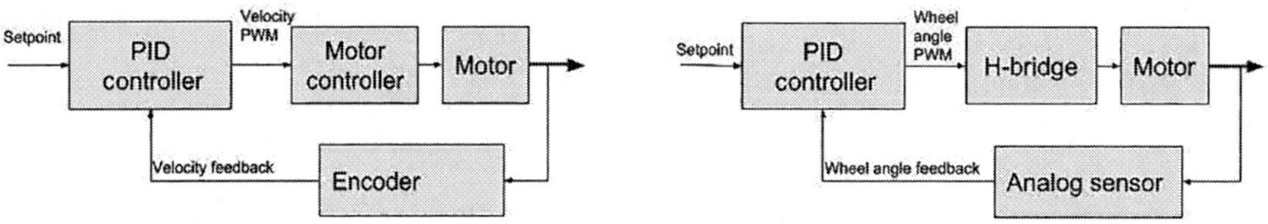

Fig. 5. Steering and main motor control diagram.

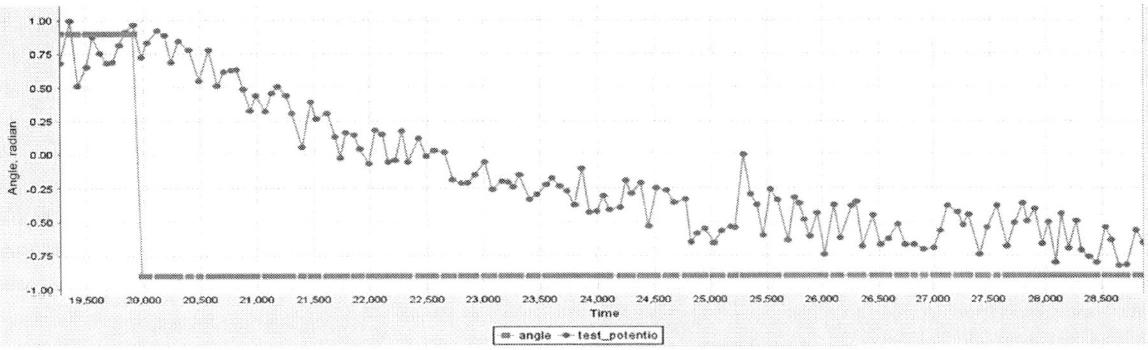

Fig. 6. PD tuning test results.

steering axle in 8.5 seconds, which is a satisfactory result for the research if we take into account the fact that the experiment was done on a standing vehicle.

7. SOFTWARE ARCHITECTURE

The high-level software architecture of the system, shown in Fig. 7, is based on the Robot Operating System (ROS). The reasons behind this decision were open-source drivers for multiple sensors and simplicity of integration of third-party software like Autoware, Yolo and multiple device drivers. According to the current architecture, the mobile robot takes several inputs from sensors. The global positioning system (GPS) is used for localization and path following, which is defined by a human. 2D (LiDAR), and camera inputs are used for obstacles' detection and safety. Output commands are steering angle, brake and linear velocity, which are sent to the low-level controllers over the UDP messages. Main software for computation of current architecture is Autoware [10], which is an open source library for self-driving cars and thus has many advanced software capabilities like lane following, obstacle avoidance, traffic light detection, lane detection etc.

The ROS platform itself is based on high modularity and scalability due to its master/slave architecture. ROS communication protocol is based on the publish-subscribe method and therefore it allows us to use external libraries and run separate individual nodes that will easily interact with each other even on multiple platforms. The ROS is a middleware and operates well on multiple cross platforms however software architecture described in this section does not use ROS on lower level controllers due to its lack of real-time capabilities. To merge ROS and lower-level controllers, software bridge was built, which converts custom ROS messages to the UDP messages. Modularity of this particular high-level software architecture is mainly the result of key principles of ROS and its approach of implementing multiple software libraries as building blocks of the primary product.

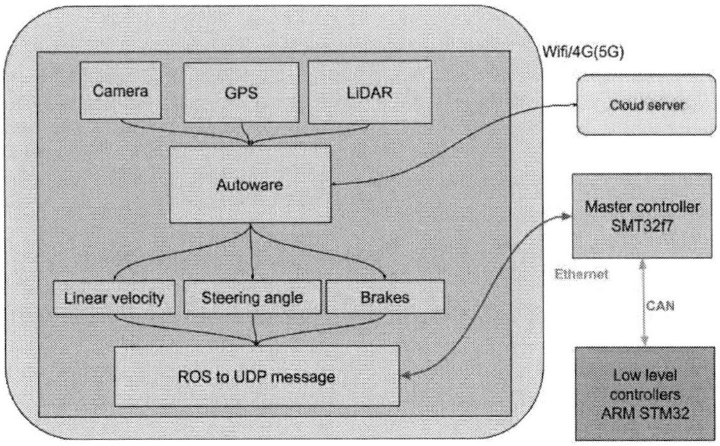

Fig. 7. Software architecture and message flow.

8. CONCLUSIONS

A modular architecture was proposed in this study to achieve flexibility and fast implementation process for off-road unmanned ground vehicles (UGVs). The architecture proposes hierarchical hardware and software structure and in particular, three-level hardware and two-level software architecture, which can be adapted according to the target platform specifics. System design and modular concept was implemented by taking into account an early stage mechatronic methodology proposed by [11]. The implementation example relies on the off-road universal robot platform UKU described in the first part of the paper. In addition, another similar robot platform is considered to reach comparable results of implementation. Both UGVs have the same size and similar purpose but they are different in their power and locomotion concept. The proposed concept for the hardware controller was evaluated in more detail using a test drive of UKU. Results showed that the proposed architecture guarantees the main driving requirements achieving stable navigation in dynamic and unknown environments, having fast and flexible implementation at the same time. The experiment of the implementation of proposed modular control system for UKU confirms that the proposed modular architecture can be easily implemented for similar unmanned robots, e.g. Agronaut. The future works include advancements in modular architecture, in particular, the high-level control algorithms including AI-based mission planning and navigation.

ACKNOWLEDGEMENTS

The research was supported by the European Commission through the H2020 project Finest Twins (grant No. 856602). The publication costs of this article were covered by the Estonian Academy of Sciences.

REFERENCES

1. Saska, M., et al. Navigation, localization and stabilization of formations of unmanned aerial and ground vehicles. In *Proceedings of the International Conference on Unmanned Aircraft Systems (ICUAS), Atlanta, GA, USA, May 28–31, 2013*. IEEE, 2013. https://ieeexplore.ieee.org/document/6564767
2. Ahmadzadeh, H., Masehian, E., and Asadpour, M. Modular Robotic Systems: Characteristics and Applications. *J. Intell. Rob. Syst.*, 2016, **81**(3–4), 317–357.
3. Wang, Y., et al. Building Unmanned Plant Factory with Modular Robotic Manipulation and Logistics Systems. In *Recent Developments in Intelligent Computing, Communication and Devices. Advances in Intelligent Systems and Computing, Vol. 752* (Patnaik, S. and Jain, V., eds). Springer, Singapore, 2019, 11–19.
4. Rassõlkin, A., Sell, R., and Leier, M. Development Case Study of the First Estonian Self-Driving Car, ISEAUTO. *Electr. Control Commun. Eng.*, 2018, **14**(1), 81–88.
5. Sell, R., Leier, M., Rassõlkin, A., and Ernits, J.-P. Self-driving car ISEAUTO for research and education. In *Proceedings of the 19th International Conference on Research and Education in Mechatronics (REM), Delft, Netherlands, June 7–8, 2018*. IEEE, 2018. https://ieeexplore.ieee.org/document/8421793
6. Rassõlkin, A., Hõimoja, H., and Teemets, R. Energy saving possibilities in the industrial robot IRB 1600 control. In *Proceedings of the 7th International Conference-Workshop Compatibility and Power Electronics (CPE), Tallinn, Estonia, June 1–3, 2011*. IEEE, 2011. https://ieeexplore.ieee.org/document/5942236
7. Väljaots, E., Sell, R., and Rimasauskas, M. Unmanned Ground Vehicle Energy Efficiency Validation in Territory Surveillance Mission. *Solid State Phenom.*, 2016, **251**, 164–170.
8. Väljaots, E., Lehiste, H., Kiik, M., and Leemet, T. Soil sampling automation using mobile robotic platform. *Agron. Res.*, 2018, **16**(3), 917–922.
9. Rassõlkin, A. and Vodovozov, V. A test bench to study propulsion drives of electric vehicles. In *Proceedings of the 8th International Conference-Workshop Compatibility and Power Electronics (CPE), Ljubljana, Slovenia, June 5–7, 2013*. IEEE, 2013. https://ieeexplore.ieee.org/ document/6601169
10. Kato, S., et al. Autoware on board: enabling autonomous vehicles with embedded systems. In *Proceedings of the 9th ACM/IEEE International Conference on Cyber-Physical Systems (ICCPS), Porto, Portugal, April 11–13, 2018*. IEEE, 2018. https://ieeexplore.ieee.org/document/8443742
11. Sell, R., Coatanéa, E., and Christophe, F. Important aspects of early design in mechatronic. In *Proceedings of the 6th International Conference of DAAAM Baltic, Industrial Engineering, Tallinn, Estonia, April 24–26, 2008*, 177–182. http://innomet.ttu.ee/daaam08/Online/Design%20Engineering/Coatanea-Sell.pdf

Mobiilsete robotite modulaarne arhitektuur

Raivo Sell, Eero Väljaots, Tengiz Pataraia ja Ehsan Malayjerdi

Modulaarne arhitektuur on võtmetähtsusega mobiilsete robotite arendusel, kui tegemist on multiotstarbelise robotsõidukiga. Artiklis on tutvustatud modulaarsuse kontseptsiooni keskklassi mobiilsele robotile, kus peamine fookus on tarkvaraline modulaarsus, iseorganiseeruvus koos lihtsalt ühendatavate lisaseadmetega, andurite väljundite kombineerimine ja riistvaraline modulaarsus. Väljatöötatud kolmetasandiline tarkvara raamistik sisaldab vahevara ja seda on rakendatud kahe erineva mobiilse roboti juhtsüsteemides.

Proceedings of the Estonian Academy of Sciences,
2019, **68**, 4, 401–406
https://doi.org/10.3176/proc.2019.4.09
Available online at www.eap.ee/proceedings

MANUFACTURING
ENGINEERING

Production monitoring system development for manufacturing processes of photovoltaic modules

Pavel Tšukrejev*, Kaarel Kruuser, and Kristo Karjust

Department of Mechanical and Industrial Engineering of Mechanical and Industrial Engineering, Tallinn University of Technology, Ehitajate tee 5, 19086 Tallinn, Estonia

Received 12 April 2019, accepted 14 May 2019, available online 19 November 2019

Abstract. The main goal of the study is to develop the concept of the real time monitoring system for manufacturing processes of photovoltaic modules. Driven by reduced costs the solar power generation is growing rapidly and increases the number of photovoltaic modules being delivered to the customers. Automatic and early detection of defects/shortcomings reduces production costs and increases productivity. This paper is focused on detecting main issues, influencing performance of the manufacturing process of photovoltaic modules. Parameters that need to be monitored will be proposed in order to eliminate the faults. The main impact factors are analysed.

Key words: solar modules, photovoltaic process, real time monitoring process, smart sensors.

1. INTRODUCTION

The current study focuses on finding possibilities of monitoring the processes inside the photovoltaic modules during manufacturing process to see how changing some parameters can influence the quality of modules and how to use this information to make further predictions. The biggest challenges were real-time measurements of different parameters inside the module without destroying it.

2. GENERAL STRUCTURE AND MANUFACTURING PROCESS OF PHOTOVOLTAIC MODULES

The burning of fossil fuels has caused big part of atmospheric pollution and the greenhouse effect in general [1]. According to [2], the growth of photovoltaic (PV) capacity of the European countries (including Turkey) in 2017 was 28% higher compared to the year 2016. The main technology for manufacturing solar modules, with market share over 90%, is the silicon-based photovoltaic cells [3,4]. Fraunhofer ISE report notes that the efficiency of the silicon cells influences the laboratory efficiency depending on the used technology. Efficiency increases in case of monocrystalline or polycrystalline silicone 26.7% and 22.3%, respectively. The market efficiency is about 21.5%. Additionally, there are also other different perspective materials that can be used. One option is to use perovskite instead of crystalline silicon inside the solar cells, as it is considered to be cheaper material reaching the efficiency of 22.1% [5]. Another example of the future material that can be used as a solar cell is the black silicon, which has significant quantum efficiency (QE) over 90% [6]. The solar modules with silicon cells are consisting of different layers, composing a sandwich structure [7] shown in Fig. 1.

* Corresponding author, pavel.tsukrejev@taltech.ee

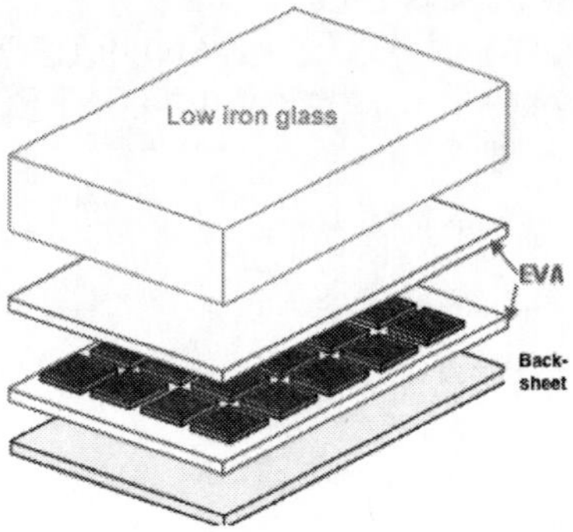

Fig. 1. Layers of solar module [8].

This paper addresses the manufacturing process of photovoltaic modules with silicon-based cells and focuses on the lamination part of it. The lamination step is essential for the manufacturing process. After lamination nothing can be changed as the whole module is already encapsulated.

3. MAIN ISSUES IN PHOTOVOLTAIC MODULES MANUFACTURING PROCESSES

As noted in [9,10], the major faults in solar modules are caused by wrong design or troubles during production. Main faults are as follows: air bubbles inside the modules; broken cells; micro cracks; hot spots; potential induced degradation (PID); snail trails. The photovoltaic modules production issues are analysed and the results are brought out in Table 1. It is explained how those types of issues and faults appear inside the module and what are the reasons. Certain suggestions, how it is possible to overcome these issues, are proposed in the last column of Table 1.

Understanding the nature of those faults helps to follow the set of selected parameters in order to prevent the issues listed above. Most of the issues in photovoltaic modules appear during lamination process. This process is very important for successful production of modules but very complicated to be tuned. Setting depends on used materials. Laminator has to perform

Table 1. Different issues in photovoltaic modules

Issue	How does it look like in the product	Why is it bad?	What is causing it?	How to avoid it?
Air bubbles	Bubbles are laminated inside the module – in different spots.	Bubbles can cause delamination. Impossible to repair.	1. Trapped air. 2. Fumes of EVA (ethylene-vinyl acetate). 3. Cooling process [9].	Fine tuning of lamination cycle is the only possibility to avoid bubbles.
Broken cells	Cells are broken inside the laminated module. Either there is a crack or a whole piece of cell is apart.	Can impact power performance. Cannot be repaired.	1. Micro cracks in cells. 2. Transportation issues. 3. Production handling. 4. Issues in production process itself.	Ensuring that handling during manufacturing and processing would not damage cells.
Micro cracks	Not visible to the eye. The crack does not penetrate the whole cell (ca 200 micrometers).	Can impact performance of cell. Risk of a real crack. Growing due to the weather influence.	1. Transportation issues. 2. Supplier ships faulty cells. 3. Wrong handling during production.	Making electroluminescent picture of every single cell that is going to be used in production.
Hot spots	The burnt spot under the glass on the surface of cell.	Spots that become warmer inside the module, leading to short circuit.	Issues in soldering the ribbon to cell. Structural defects of cell.	Ensuring the quality of soldering interconnectors.
PID	Cannot be visually determined.	Accelerated aging of modules. Solar modules could lose performance.	Potential difference between solar module and the earthing.	Improving design of the module and the choice of materials used.
Snail trails	"Browning" of contact fingers of cell in a form of trails. Appears after some years.	Appearance of module. Can cause loss in power output.	Moisture penetrating the module causing oxidation of the components of cells on the trails of micro cracks.	Avoiding different types of cracks in cells.

the whole cycle (up to 25 minutes) in order to produce a module. During lamination the chamber is closed and it is not possible to see if something is going wrong there. After every change in the program it is important to wait the whole cycle. In case of wrong settings modules will be scrapped.

There are different steps in lamination process with different temperatures and pressures. The encapsulant material EVA becomes liquid during lamination. This may influence cells to float inside and cause undesirable displacements. During the lamination EVA transforms and forms a polymeric connection inside itself.

3.1. Measurable parameters

The study of lamination process allows selecting the parameters that need to be tracked:

- *pressure (input)* – essential during lamination process (impact on the quality of encapsulant and appearing of air bubbles);
- *temperature (input)* – main parameter for photovoltaic module lamination. Impacting the quality of encapsulant;
- *duration (input)* – lamination time. Influences the content of EVA gel and bubbles inside the laminate;
- *displacements (output)* – is essential to understand if module's structure is right. Finding displacements which are not visible to the eye and understanding why they occurred helps to improve modules in general;
- *content of EVA gel (output)* – important parameter to be followed but hard to be tracked by sensors in real time, as it is determined during chemical processes;
- *cracks (output)* – finding different types of cracks before laminating process allows to replace the cell;
- *measured temperature (temporary output)* – installation of additional sensors to collect data about the temperature should be considered as it could help making decisions how to measure temperature in a right way. Figure 3 shows the experimental setup of photovoltaic module with thermal sensors installed on the top of the module.

In order to track those parameters, the main idea is to embed a sensor inside the photovoltaic module body in the stage of production. Embedded sensors inside the composite materials was studied by Herranen et al., in [12] and the optimal shape for sensor's protective housing was proposed. The optimization procedure based on combining hybrid genetic algorithm (HGA), artificial neural networks (ANN) and reduced-order models (ROMs) [13–17], was adapted for optimal design of housing of the electronic component.

4. DEVELOPING THE CONCEPT OF MONITORING METHOD FOR PRODUCTION OF PHOTOVOLTAIC MODULES

The real-life production monitoring stands for continuous checking of parameters from the manufacturing object [18, 19]. This could be a powerful tool for measuring necessary parameters and understanding what is happening inside the product during the manufacturing process. There is possibility to use different types of sensors, which could be embedded inside the photovoltaic module, e.g., RFID (radio-frequency identification) sensors [11].

Before monitoring the manufacturing process of photovoltaic modules, it is necessary to identify important parameters that need to be monitored during production process. In other words, key performance indicators (KPIs) will have to be defined for using enterprise analyses model (EAM) [20–24].

The KPIs should be related to two main fields: jobs and courses and cover different areas like machinery, materials, processes, employees, facility, products, production order, etc.

4.1. Basic concept of monitoring system for photovoltaic modules

There is a number of approaches and requirements for the production monitoring systems. Let us proceed with the basic concept of production monitoring system developed by the workgroup in [18] and shown in Fig. 2.

This concept is proposed for particular case study with additional measurable parameters like pressure, temperature, and duration. This system can be adapted to the manufacturing process of photovoltaic modules with some changes. The concept proposed for production monitoring system integrates the following five main modules: data collection, analysis, visualization, storage, and security. The sensor system is collecting different data (ambient temperature, pressure, duration, displacements, and electroluminescence) from workstations. The storing, computations and analyses are made in the server or cloud side; and integration between the monitoring system and enterprise resource planning (ERP) system is developed. The collected information is sent to the ERP system where it is visualized.

4.2. Tests of temperature measurement

As noted above the lamination temperature has impact on quality of encapsulant and can be considered as key factor in lamination process. In order to validate the

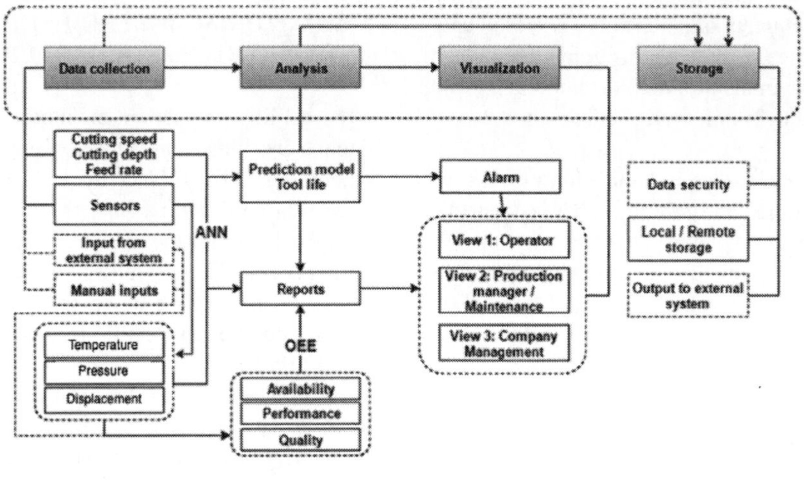

Fig. 2. Concept of the production monitoring system [18].

concept of measuring parameters inside the laminating chamber the temperature measurement was performed. The thermometer TES 1312A with two wired temperature sensors was employed. Sensors were installed on the back of solar module attached to the backsheet. Placing of thermal sensors can be seen in Fig. 3.

There were three tests performed: (1) usual lamination program with temperature of 144 °C; (2) program with increased temperature of 150 °C; (3) program with reduced temperature of 130 °C. The results of tests and gathered temperature data are represented in Table 2 and Fig. 4. Two measured values of temperature (Temp. 1 and Temp. 2 in Table 2) can be considered as an output from particular temperature test. The temperature was checked and registered with time stamp and values of

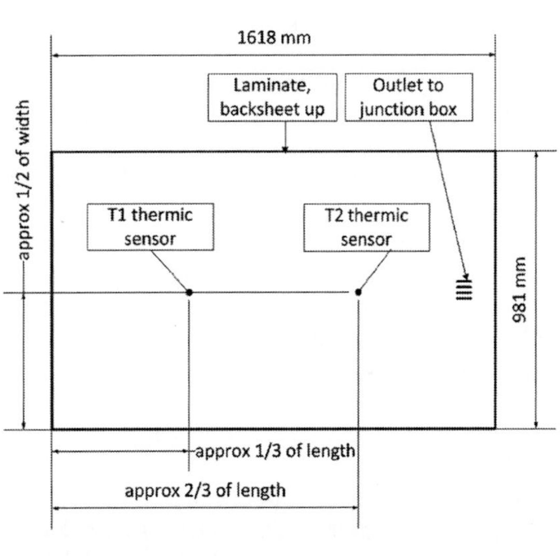

Fig. 3. Schematic representation of experimental setup.

measured temperatures. The set value of the temperature is known from laminator receipt and is given also in Table 2. "Machine values" are the values that are either set by receipt or measured by machine, not by experimental setup.

Gathered results show that there is a difference in set and real temperatures. The reason is the fact that heating element of the machine with temperature sensor is located underneath the heating plate. The module is located on the heating plate and the thermal sensor used in the test was set on the top of the module, so the whole structure of the module is between two sensors. Surface temperature of the module never reached program receipt value by the end of lamination cycle in any of tests. One important remark is that wired sensors were used in experimental set-up and wires had left traces on the surface of the backsheet. This means that particular module was defined faulty and was scrapped. Wireless solution or other type of wires should be used.

In Fig. 4, temperature measurement results from Table 2 are presented according to performed test number (the first number) and sensor position (decimal number). Measurements were performed five times.

5. CONCLUSIONS

It is essential to understand, what is happening inside the modules during production as well as to react fast to aoid faults in products. This kind of data gathering and reacting to the results of analysis will help to improve overall quality of products as well as understand what is happening inside the modules during different stages of production.

Table 2. Results of the tests

Measure-ment	Time from start	Machine values				Measured	
		Set temp., °C	Measured temp. value, °C	Set pressure, mbar	Meaured pressure, mbar	Temp. 1, °C	Temp. 2, °C
1.1	2 min 47 s	144	143	–1000	–1024	60.3	68.7
1.2	6 min 42 s	143	142	–300	–303	95.6	99.3
1.3	10 min 4 s	143	143	–300	–293	117.8	119.1
1.4	14 min 22 s	143	143	–300	–289	130.3	130.7
1.5	14 min 54 s	144	143	0	–2	131.1	130.9
2.1	2 min 44 s	150	149	–1000	–1006	51.7	57.5
2.2	6 min 16 s	150	148	–300	–310	80.9	87.2
2.3	10 min 1 s	150	150	–300	–300	117.4	118.1
2.4	14 min 31 s	150	150	–300	–307	134.5	134.4
2.5	15 min 00 s	150	150	0	7	134.4	135.8
3.1	2 min 34 s	130	130	–1000	–1013	36.0	32.9
3.2	6 min 01 s	130	129	–300	–361	71.4	74.7
3.3	10 min 5 s	130	130	–300	–320	100.2	103.3
3.4	14 min 35 s	130	130	–300	–300	118.7	120.5
3.5	15 min 00 s	150	150	0	7	134.4	135.8

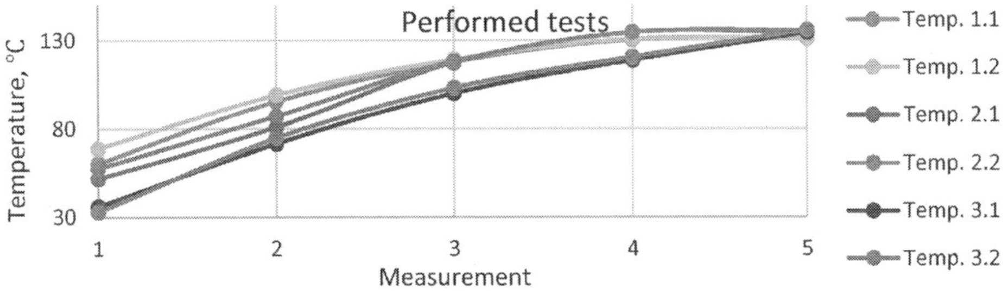

Fig. 4. Temperature test results in different positions and input data.

In the current paper, the main issues and their sources have been pointed out. The concept for production monitoring system is proposed. The real time tests for measuring set temperature and temperatures in two additional nodes are performed and the results are analysed. Further studies will be performed regarding embedding sensors into solar modules.

ACKNOWLEDGEMENTS

This research was supported by the Estonian Research Council (grant PUT1300); the Estonian Centre of Excellence in Zero Energy and Resource Efficient Smart Buildings and Districts; ZEBE, grant TK146 funded by the European Regional Development Fund. The publication costs of this article were covered by the Estonian Academy of Sciences.

REFERENCES

1. Graus, W. and Worrell, E. Trend in efficiency and capacity of fossil power generation in the EU. *Energy Policy*, 2009, 37(6), 2147–2160.
2. Clover, I. and Tsagas, I. Europe added 8.61 GW of solar in 2017 with Turkey taking top spot. *PV Mag.*, 2018. https://www.pv-magazine.com/2018/02/09/europe-added-8-61-gw-of-solar-in-2017-with-turkey-taking-top-spot/
3. Silvestre, S., Tahri, A., Tahri, F., Benlebna, S., and Chouder, A. Evaluation of the performance and degradation of crystalline silicon-based photovoltaic modules in the Saharan environment. *Energy*, 2018, **152**, 57–63.
4. Photovoltaics report. Fraunhofer Institute for Solar Energy Systems, ISE with support of PSE Conferences & Cosulting GmBH. 2018 (accessed 2018-02-26). https://www.ise.fraunhofer.de/content/dam/ise/de/documents/publications/studies/Photovoltaics-Report.pdf

5. Shi, Z. and Jayatissa, A. H. Perovskites-Based Solar Cells: A Review of Recent Progress, Materials and Processing Methods. *Materials (Basel)*, 2018, **11**(5), 729.
6. Juntunen, M. A., Heinonen, J., Vähänissi, V., Repo, P., Valluru, D., and Savin, H. Near-unity quantum efficiency of broadband black silicon photodiodes with an induced junction. *Nat. Photonics*, 2016, **10** (12), 777–781.
7. Abdelkader, E. A., Mahrane, A., Moussa, F. Y., and Boukennous, Y. Solar Module Fabrication. *Int. J. Photoenergy*, 2007. http://downloads.hindawi.com/journals/ijp/2007/027610.pdf https://www.researchgate.net/publication/26488380_Solar_Module_Fabrication
8. Honsberg, C. and Bowden, S. Module Materials. http://www.pveducation.org/pvcdrom/modules/module-materials (accessed 2018-12-12).
9. Perret-Aebi, L.-E., Li, H.-Y., Théron, R., Roeder, G., Luo, Y., Turlings, T., Lange, R. F. M., and Ballif, C. Insights on EVA lamination process: Where do the Bubbles Come From? In Proceedings of the 25th European Photovoltaic Solar Energy Conference and Exhibition /5th World Conference on Photovoltaic Energy Conversion, Valencia, Spain, September 6–10, 2010. https://www.researchgate.net/publication/47806645_Insights_on_EVA_lamination_process_Where_do_the_bubbles_come_from
10. Meyer, S., Turek, M., Großer, S., Manke, T., and Hagendorf, C. Understanding snail trails on PV modules. *PV Tech*, 2017, **10**, 60–64.
11. Zhou, S., Ling, W., and Peng, Z. An RFID-based Remote Monitoring System for Enterprise Internal Production Management. *Int. J. Adv. Manuf. Technol.*, 2007, **33**(7–8), 837–844.
12. Herranen, H. *Design Optimization of Smart Composite Structures with Embedded Devices*. TUT Press, Tallinn, 2014.
13. Herranen, H., Majak, J., Tšukrejev, P., Karjust, K., and Märtens, O. Design and manufacturing of composite laminates with structural health monitoring capabilities. *Procedia CIRP*, 2018, **72**, 647–652.
14. Aruniit, A., Kers, J., Goljandin, D., Saarna, M., Tall, K., Majak, J., and Herranen, H. Particulate Filled Composite Plastic Materials from Recycled Glass Fibre Reinforced Plastics. *Mat. Science*, 2011, **17**(3), 1–6.
15. Majak, J., Pohlak, M., Eerme, M., and Velsker, T. Design of car frontal protection system using neural networks and genetic algorithm. *Mechanika*, 2012, **18**(4), 453–460.
16. Lellep, J. and Majak, J. Nonlinear constitutive behavior of orthotropic materials. *Mech. Compos. Mater.*, 2000, **36**(4), 261–266.
17. Karjust, K., Pohlak, M., and Majak, J. Technology Route Planning of Large Composite Parts. *Int. J. Mater. Form.*, 2010, **3**(1), 631–634.
18. Eiskop, T., Snatkin, A., and Karjust, K. Production monitoring system with predictive functionality. *J. Eng. Sci. Technol.*, 2017, **12**(9), 2410–2425.
19. Prasad, H. C. S., Rao, P. S., Gopala Krishna, B., and Ahluwalia, A. Productivity enhancement through production monitoring system. *Int. J. Appl. Bus. Econ. Res.*, 2016, **14**(3), 1755–1767.
20. Kaganski, S., Majak, J., Karjust, K., and Toompalu, S. Implementation of key performance indicators selection model as part of the Enterprise Analysis Model. *Procedia CIRP*, 2017, **63**, 283–288.
21. Sell, R., Coatanéa, E., and Christophe, F. Important aspects of early design in mechatronic. In *Proceedings of the 6th International DAAAM Baltic Conference "Industrial Engineering"*, Tallinn, Estonia, April 24–26, 2008. http://innomet.ttu.ee/daaam08/Online/Design%20Engineering/Coatanea-Sell.pdf
22. Sell, R., Aryassov, G., Petritshenko, A., and Kaeeli, M. Kinematics and dynamics of configurable wheel-leg. In *Proceedings of the 8th International DAAAM Baltic Conference "Industrial Engineering"*, Tallinn, Estonia, April 19–21, 2012, 345–350. http://innomet.ttu.ee/daaam_publications/2012/sell.pdf
23. Polyantchikov, I., Shevtshenko, E., Karaulova, T., Kangilaski, T., and Camarinha-Matos, L. M. Virtual enterprise formation in the context of a sustainable partner network. *Ind. Manage. Data Syst.*, 2017, **117**(7), 1446–1468.
24. Mahmood, K., Karaulova, T., Otto, T., and Shevtshenko, E. Performance Analysis of a Flexible Manufacturing System (FMS). *Procedia CIRP*, 2017, **63**, 424–429.

Tootmise seiresüsteemi arendamine fotoelektriliste moodulite tootmisprotsesside jaoks

Pavel Tšukrejev, Kaarel Kruuser ja Kristo Karjust

Fotoelektriliste moodulite tootmise seiresüsteemi väljaarendamiseks on läbi viidud uuring põhilistest kvaliteediprobleemidest, mis võivad tekkida moodulite valmistamise käigus. Uuringust selgus, et enamik probleeme tekib lamineerimise etapi jooksul. Tehti kindlaks põhjused, mis viivad kvaliteediprobleemide tekkimiseni. Põhjuste alusel on koostatud loend parameetritest ja mõjufaktoritest, mida on vaja lamineerimisel jälgida. Üldine tööpõhimõte seiresüsteemi arendamisel on integreerida moodulisse andurid, teha mõõtmised ja edastada tulemused serverisse. Saadud andmete põhjal saab teostada analüüsi, prognoosi ja visualiseerimist.

Proceedings of the Estonian Academy of Sciences,
2019, **68**, 4, 407–412
https://doi.org/10.3176/proc.2019.4.10
Available online at www.eap.ee/proceedings

MANUFACTURING ENGINEERING

Implementation of a knowledge-based manufacturing on the example of Sumar Tools OÜ

Kaarel Kruuser[a*], Jüri Riives[a], Pavel Tšukrejev[a], and Indrek Kiolein[b]

[a] Department of Mechanical and Industrial Engineering, Tallinn University of Technology, Ehitajate tee 5, 19086 Tallinn, Estonia
[b] Pro-Step OÜ, Tööstuse 47b-7, 10416 Tallinn, Estonia

Received 12 April 2019, accepted 14 May 2019, available online 19 November 2019

Abstract. Higher productivity of an organization is based on modern manufacturing systems and digitizing the manufacturing processes. The problem is how to match together technological capabilities of machine tools (machining centres) with the need to manufacture a product; and how to increase the efficiency of process planning procedures. Injection molds are the products with many different features, with high quality requirements, and variable geometrical parameters. The problem is sophisticated because most of the products are unique and do not repeat. In order to be able to respond quickly changeable situations where sometimes more than 4 different programs are needed on a daily basis, manufacturing systems will have to become more autonomous. For this purpose, it is necessary to take advantages of the possibilities of modern production programs by linking them with know-how of engineers. Manufacturing operations management (MOM) and machine learning are the tools used for developing knowledge-based manufacturing solutions in the company. Framework has been developed to gather data in the company and develop the rules based on data models and information analysis. The designed solutions for feature-based machining (FBM) give high quality technological solutions and increase the efficiency of machining process by computer numerical control (CNC) programming. Product manufacturing information (PMI) model is in development together with integration of the machine learning system.

Key words: feature-based machining, knowledge-based manufacturing.

1. INTRODUCTION

Sumar Tools OÜ is a company that manufactures sophisticated plastic injection molds. The production of plastic injection molds requires high level knowledge in manufacturing technology and formulation. In order to be capable of producing plastic injection molds there is a need for accurate and high-end machine park. The company mainly uses milling; also turning centres, electrical discharge machining (EDM); also wire erosion and grinding machines to manufacture injection molds. The company's machine park also includes an innovative and high-level flexible manufacturing system. The correct use of the system ensures high productivity. The flexible manufacturing system has been integrated with one 5-axis and one 7-axis machining centre. Machining centres are capable of processing technologically complex and high precision products that require 5- and 7-axis processing technology.

It is difficult to use a flexible manufacturing system (FMS) in the company where each product is different and there is no series production. The key question is how to keep the production system cost-effective. Also, it is necessary to find the solution how to ensure maximum productivity and keep the system sustainable.

One possible solution is to find similar product lines and cluster them. To do this, we have to analyse the product portfolio and divide products into groups

* Corresponding author, kaarel.kruuser@gmail.com

according to the similar features. Grouping products by the features allows to develop optimal technology for the processing features. This type of manufacturing technology is called feature-based machining (FBM).

FBM is an automated programming function that eliminates manual data entry by reading in the copious design intent information that is embedded in solid models by CAD programs [1].

2. THEORETICAL BACKGROUND

Sumar Tools OÜ is using Siemens NX [2,3] design and manufacturing software (CAD CAM). This software is flexible and powerful integrated solution, which helps to deliver production faster and more efficiently. Siemens NX software enables using feature-based machining technology which allows more efficient design and manufacturing preparation.

Without machining feature the program generation is manual and the cutting path is based on the given coordinate system. Programming will take longer time and machining efficiency is significantly lower compared to the FBM. Feature-based machining can be implemented in three ways as shown in Fig. 1.

Design by the feature [5] is a technique by which the geometrical features are described and documented in a certain data-base. For instance, to define a hole, designer can specify radius, depth and parameters of location. Described features are creating the possibilities for their classification, to define the rules for their machining, and developing the similarity assessment algorithms [6].

The second approach is developed by means of pattern recognition. This method is largely used in computer-aided manufacturing (CAM) software. However, generating NC program by commercial CAM software will result in longer part program besides utilizing a large memory space in CNC machine controller [7].

The last approach is human-assisted feature recognition based on the NC program generation. This leads towards parametric programming implementation in feature-based machining [8].

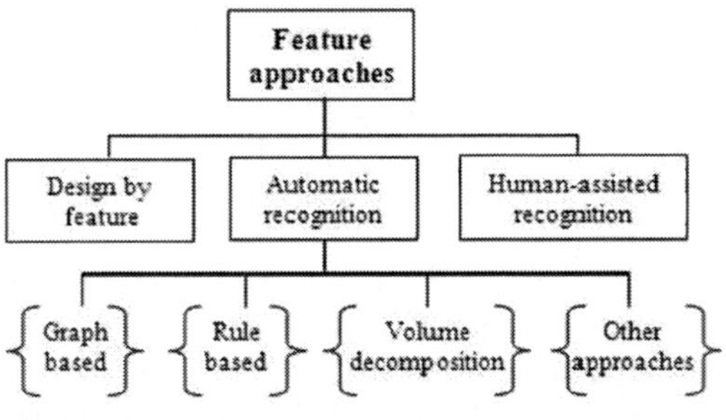

Fig. 1. Feature approaches classification [4].

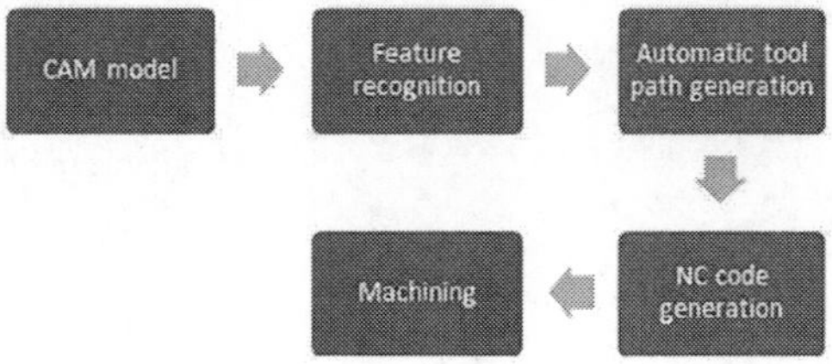

Fig. 2. Modern path from model to machining.

FBM technology enables automation of preparation process of the manufacturing technology. Creation of CNC programs has risen to the new level by making manufacturing faster, more flexible and cost effective (see Fig. 2). The technology is based on the feature elements which are found from the 3D model of the product. This technology is developed to identify features and compile a technology for manufacturing elements.

Siemens NX uses information about the product and manufacturing from the 3D model (PMI), e.g. machining tolerance and surface quality information. PMI information gathered from the model is analysed and processed. The processing technology is selected based on the information.

To automate the workflow with feature-based machining, different databases will have to be developed. Databases are created by collecting information and knowhow from the company. At first, main rules are necessary to form a foundation for feature recognition. The movement of collected data is shown in Fig. 3.

3. IMPLEMENTATION OF FEATURE-BASED METHODOLOGY IN SUMAR TOOLS OÜ

The implementation of FBM technology at the company is a labor-intensive process, which however, should lead to a considerable time-saving and efficiency in preparation of the manufacturing process. Key performance indicators (KPIs) will have to be determined for separate processes and steps to analyse the implementation process [9,10]. Feature-based machining, product life-cycle management (PLM) and real-time monitoring systems of production processes should be integrated to reach the next productivity level [11,12].

In traditional manufacturing processes preparation of CNC programs for the machining centres makes almost 75% of the CAM engineer's workload. At the moment the company has two CAM engineers. One of them is committed to prepare the production technology for the

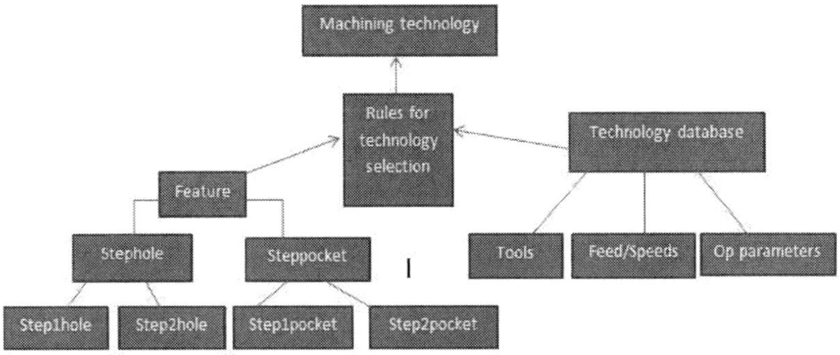

Fig. 3. Data movement in FBM system.

plate-type products. The first step in the implementation is to classify and group the features of the products. In this case, the products are divided into three groups:

(1) Group 1 – products with low precision feature elements;
(2) Group 2 – products with high precision feature elements;
(3) Group 3 – products with high precision features and 3D surfaces.

Products with low precision features include details where dimensional tolerances are not high. As a result, it is possible to process features with one operation, without stocks, ensuring that the design elements meet the precision requirements.

Products with high precision features include details where dimensional tolerances are high. This means that the elements must be processed in number of different operations to ensure that the features meet the precision requirements. In case of more precise tolerances, the product may move to different processing stations, which results in the FBM-generated program assigning workflow operations. For the sake of technological correctness, work operations are divided into groups as follows: rough machining, semi-finish machining, finish machining.

Products with high precision feature elements and 3D surfaces with high tolerance requirements add complexity to the programming of FBM. It is difficult to standardize the complex operations.

As a result of the analysis of comprehensive portfolio of products, it is possible to create knowledge bases for the features of design. The feature elements can be divided into two larger groups: step hole and step pocket. We can divide these two main groups into sub-groups according to the number of steps (see in Fig. 4).

In the next technological development phase, the machining technology was developed for each group member, see Fig. 5. The numbers in the table show which operations in which order are selected for different

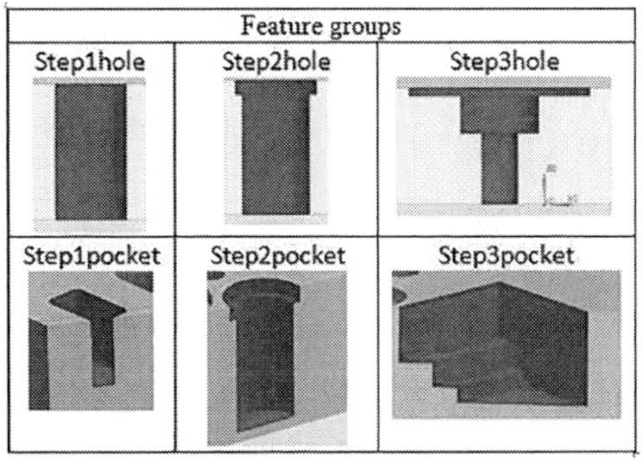

Fig. 4. Features divided into groups according to the number of steps found in the feature.

technologies. Manufacturing technology was chosen based on the working methods and capabilities of the company. For the real use of system technology, automatically generated technology must match the company's knowledge and capabilities.

To do this, the following databases must be created: tool database, database of used materials, database of cutting feed and speeds, and database of cutting operational parameters.

After the development of databases, it was necessary to create the selection rules for selection, according to which the system selects the machining operation, the tool and the corresponding cutting parameters. The PMI information is used for selecting the technology. PMI information allows decision making in the process to choose between different technologies, to make the system more flexible and more precise.

Product manufacturing information (PMI) is available from the 3D model to make the expert system work more efficiently. In case of standard technology, PMI information is available in 2D drawings. In order to avoid redundant workload of the CAM engineers and to prevent possible errors, it is necessary to involve the CAD engineer into the process in order to add PMI

Output Feature	STEP3HOLE																				
Operation set	1	2	3	4	5	6	7	8	9	10	11	12	13	14	15	16	17	18	19	20	21
↓ Input Feature ↓																					
STEP3HOLE																					
DRILL_S3H_D1	1						1	1	1	1	1	1	1	1					1	1	1
DRILL_S3H_D2	2	1					2	2	2	2	2	2	2	2					2	2	2
DRILL_S3H_D3	3	2	1	1	1	1	3	3	3	3	3	3	3	3	1	1	1	1	3	3	3
COUNTERMILL_S3H_D1		3	2	2	2	2									2	2	2	2			
COUNTERMILL_S3H_D2			3	3	3	3									3	3	3	3			
COUNTERMILL_S3H_D1_FINISH_BOT				4		4	4	4	4	4	4	4	4	4			4	4			
COUNTERMILL_S3H_D2_FINISH_BOT					4	5	5	5	5	5	5	5	5	5			5	5			
CHAMFER_S3H_CONTOUR_D1	When chamfer is needed																				
CHAMFER_S3H_CONTOUR_D2																					
CHAMFER_S3H_CONTOUR_D3																					
CHAMFER_S3H_CONTOUR_D3_REV																					
BORE_S3H_D3							6			6											
REAM_S3H_D3											6			6							
BORE_S3H_D2								6		7											
REAM_S3H_D2												6		7							
BORE_S3H_D1									6						4		6				
REAM_S3H_D1													6			4		6			
TAP_S3H_THREAD_D1																			4		
TAP_S3H_THREAD_D2																				4	
TAP_S3H_THREAD_D3																					4

Fig. 5. Selecting the machining technology according to the feature.

information to the 3D model. During the design process, the engineer works through the measuring circuits required for the functionality and operation of the product, and binds them with the 3D model using PMI tool that is designed for this purpose in the CAD system.

Using PMI in the design phase gives us valuable information about the features, accuracy, surface roughness, shape tolerances etc.

The use of PMI is the best solution so far for transferring engineers' wishes into expert system. In addition, PMI information can also be displayed in drawings, avoiding repetition of the same work.

This technology is developed to analyse PMI information, and for deciding what kind of manufacturing technology would be the best to use. Feature element class is determined based on the precision class and required surface quality. The precision feature is low if, the dimensional tolerance is greater than ± 0.1 mm, and the required roughness is greater than Ra 3.2 μm.

The precision feature is high if the dimensional tolerance is lower than ± 0.1 mm, and the required roughness is lower than Ra 3.2 μm.

In case the feature is classified as low precision feature, a simple manufacturing technology is used. The dimensions of the feature are achieved by creating only rough operations with stock left to zero.

In case of the high precision classified feature, more sophisticated technology is used. Generated operations are divided into rough, semi-finish and finish machining. Set of rules are developed for technology decision making. System is analysing tolerance grade based on PMI; and deciding requires extra operations, see in Fig. 6.

The tool database is created according to the tools used by the company. When creating a database, it is important that the tool is accurately described and the data correspond to the reality. This ensures that, when operation is created, a selected tool is correct and the usable operation is produced.

When describing the tool in the database, the tool's basic parameters are entered and a note indicating in which operation the tool was created (rough, finish) is added.

The choice of instruments of the database is based on technology, which results in selection of a tool suitable for the operation. Rules which have been developed for making suitable tool selection are shown in Fig. 7.

Tool selection is also influenced by the material of the workpiece; various instruments can be configured

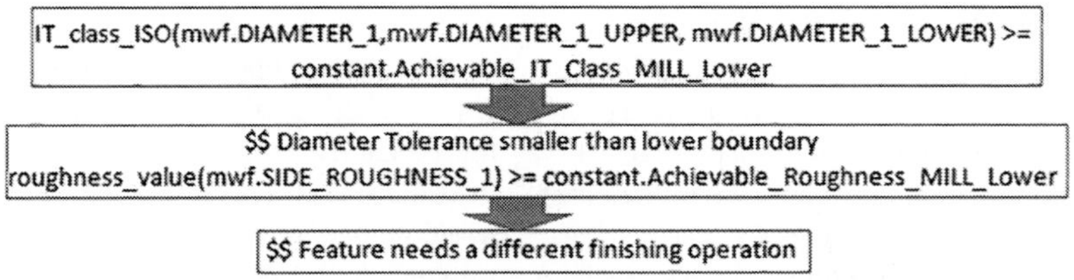

Fig. 6. Decision making rules for technology selection.

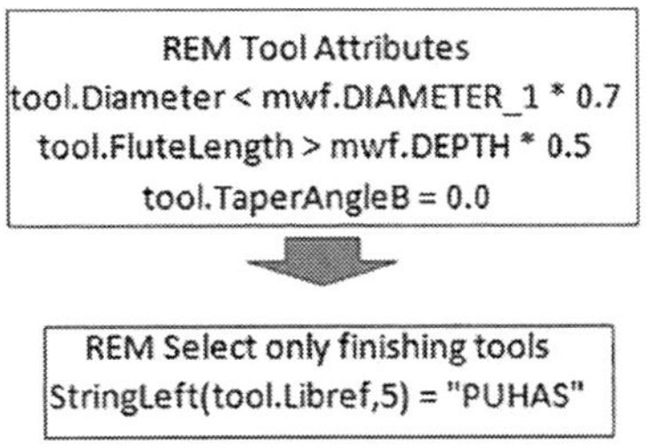

Fig. 7. Decision making rules for tool selection.

for different materials. A material database has to be prepared for the manufacturing technology. When creating a technology database, all created databases must be connected, see Fig. 8.

Adding all databases to PLM creates an expert system that can be used to prepare CNC programs for machining centres. After inserting a product into the CAM, it is necessary to identify the detail and the position of the detail in the system. Then let the system analyse and identify the features and generate work paths as shown in Fig. 9.

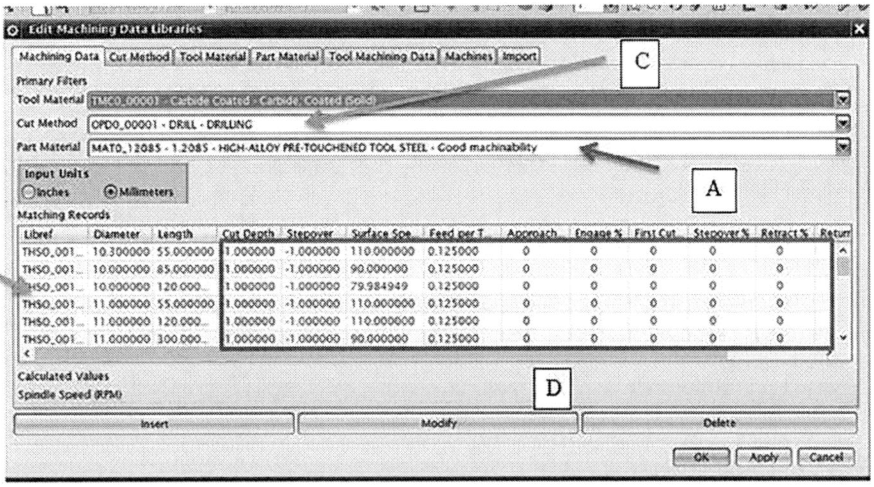

Fig. 8. Interconnection of databases. A – material, B – tools, C – operation type, D – feeds and speeds.

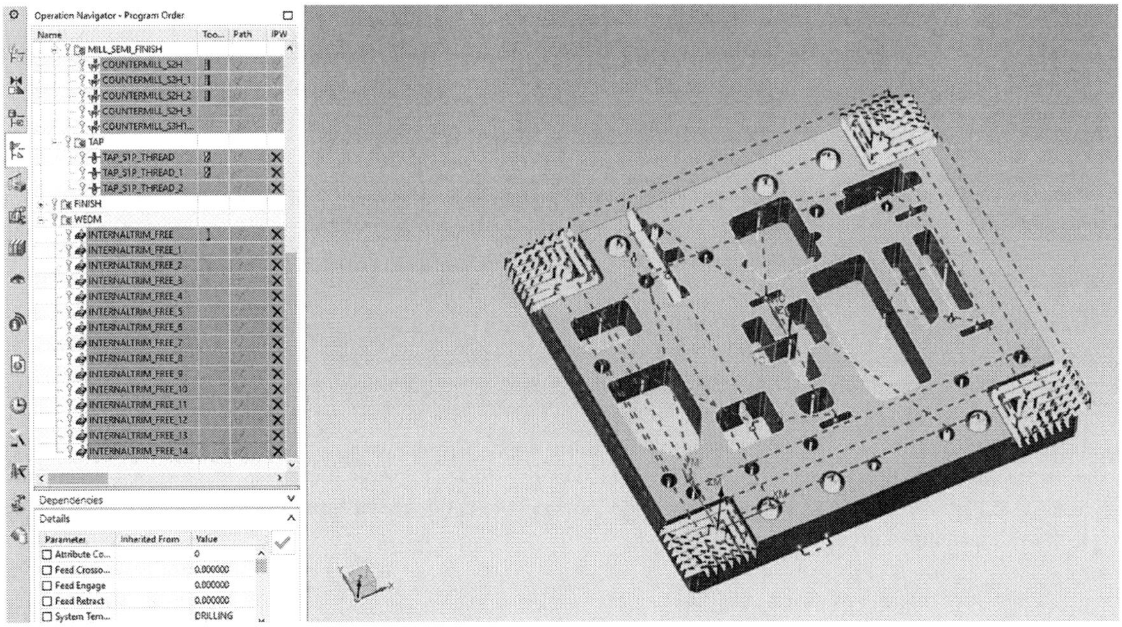

Fig. 9. Example work paths generated by the FBM.

4. CONCLUSIONS

It can be concluded from the results of implementing FBM technology and received feedback, that applying FBM technology will considerably accelerate preparation of manufacturing process. Using FBM technology can reduce almost 20% of CAM engineers' workload. Creating a production preparation expert system can help to standardize technology design. This system can be used for producing almost all kind of plate-type products which are produced in the company. Almost 95% of holes are machinable with FBM technology. This technology involves the entire manufacturing structure of the company, and the Teamcenter solution simplifies the exchange of data between different departments. Although the technology deployment process is time-consuming, the efficiency achieved is noticeable. It also allows the system to optimize technologies due to its continuous storage system.

ACKNOWLEDGEMENTS

This research was supported by the Innovative Manufacturing Engineering Systems Competence Centre IMECC (supported by Enterprise Estonia and co-financed by the European Union Regional Development Fund, project EU48685). The publication costs of this article were covered by Tallinn University of Technology and the Estonian Academy of Sciences.

REFERENCES

1. Shah, J. J., Mäntylä, M., and Nau, D. S. *Advances in Feature Based Manufacturing*. Elsevier Science B. V., Amsterdam, 1994.
2. Leu, M. C., Ghazanfari, A., and Kolan, K. NX 10 for Engineering Design. Missouri University of Science and Technology, 2016. https://web.mst.edu/~mleu/nx_manuals/nx10.pdf
3. Siemens NX software https://www.plm.automation.siemens.com/global/en/products/nx/ (accessed 2018-10-21).
4. Mokhtar, A., Tavakoli-Bina, A. T., and Houshmand, M. Approaches and challenges in machining feature-based process planning. In *Proceedings of the 4th International Conference on Digital Enterprise Technology (DET2007) Bath, UK, September 19–21, 2007*, 297–305.
5. Venkataraman, S. Integration of design by features and feature recognition. MSc thesis. Arizona State University, Arizona, 2000.
6. Cardone, A., Gupta, S. K., Deshmukh, A., and Karnik, M. Machining feature-based similarity assessment algorithms for prismatic machined parts. *Comput.-Aided Des.*, 2006, **38**(9), 954–972.
7. Al'Hapis, M., Razak, A., and Zakaria, A. A Framework for a Feature Based Machining Using Macro. *Appl. Mech. Mater.*, 2012, **110–116**, 1711–1715.
8. Djassemi, M. A parametric programming technique for efficient CNC machining operation. *Comput. Ind. Eng.*, 1998, **35**(1–2), 33–36.
9. Kaganski, S., Majak, J., Karjust, K., and Toompalu, S. Implementation of Key Performance Indicators Selection Model as Part of the Enterprise Analysis Model. *Procedia CIRP*, 2017, **63**, 283–288.
10. Paavel, M., Karjust, K., and Majak, J. Development of a product lifecycle management model based on the fuzzy analytic hierarchy process. *Proc. Est. Acad. Sci.*, 2017, **66**(3), 279–286.
11. Eiskop, T., Snatkin, A., and Karjust, K. Production monitoring system with predictive functionality. *J. Eng. Sci. Technol.*, 2017, **12**(9), 2410–2425.
12. D'Antonio, G., Segonds, F., Laverne, F., Bedolla, J. S., and Chiabert, P. A framework for manufacturing execution system deployment in an advanced additive manufacturing process. *Int. J. Prod. Lifecycle Manage.*, 2017, **10**(1), 1–19.

Teadmuspõhise tootmise juurutamine Sumar Tools OÜ-s

Kaarel Kruuser, Jüri Riives, Pavel Tšukrejev ja Indrek Kiolein

Targa tootmise juurutamine aitab tootmist efektiivsemaks muuta. Targa tootmise põhimõtte raames juurutati kujuelementidel põhinev töötlusprogrammide tegemise moodul, mis on ettevõttes kasutusel oleva tootmistarkvara Siemens NX üks lisamoodulitest. Mooduli kasutamiseks tuli analüüsida kogu ettevõtte tooteportfelli. Analüüsi tulemusel jagati tooted nendel leiduvate kujuelementide põhjal kolme gruppi. Iga kujuelemendi töötlemiseks loodi vastav tehnoloogia vastavalt 3D-mudelilt saadava PMI info analüüsi tulemusele. Lisaks tehti materjalide, tööriistade ja töötlusparameetrite andmebaasid. Nende omavaheliseks koostööks tuli luua käitumise reeglid, mis võimaldavad valida õiget töötlemise tehnoloogiat. Kujuelementidel põhineva tehnoloogia kasutuselevõtu tulemusel muutus tehnoloogia ettevalmistamine kiiremaks ja efektiivsemaks.

Proceedings of the Estonian Academy of Sciences,
2019, **68**, 4, 413–420
https://doi.org/10.3176/proc.2019.4.11
Available online at www.eap.ee/proceedings

MANUFACTURING
ENGINEERING

Methodology for implementing universal gripping solution for robot application

Mohammed Salman Azim[a*], Andrei Lobov[a], and Artem Pastukhov[b]

[a] Mechanical Engineering and Industrial Systems, Tampere University, Korkeakoulunkatu 7, P.O. Box 589, 33101 Tampere, Finland
[b] ITMO University, International Research Centre "Biotechnologies of the Third Millennium", Lomonosov Str. 9, 191002 Saint Petersburg, Russia

Received 12 April 2019, accepted 7 May 2019, available online 19 November 2019

Abstract. In recent years the affordability of robots and the progress in collaborative robotics has been of great benefit for the manufacturing industries. The repetitive, monotonous and eco-unfriendly tasks are being assigned to the robots, which can work in parallel with humans, making the tasks easier for them. Industries are frequently introducing robots on the factory floor for maximizing production. Competition on the market is motivating robot manufacturers to work out solutions where return on investment would take as little time as possible. End effector is the most important part of a robot for making specific operations. The end effector market has also grown and brought innovation in the area of grasping objects with different shapes with a single gripper. However, problems persist due to the need for a gripper, which could handle a diverse range of products for certain applications. This paper discusses an approach of handling different products with a single end effector. Selecting a gripper for a certain application takes time and effort. Universal gripping solution can provide extra benefits and save costs. Here, a methodology is proposed to design a proper universal gripping solution for a specific use case. The article is mainly focused on pick-and-place applications.

Key words: robot gripper, universal gripper, 3D printed gripper, gripper selection.

1. INTRODUCTION

Since the first inclusion of industrial robots in early 1960s, industries have welcomed robots on the factory floor due to the vast range of applications [1]. Robotic systems can be implemented starting from automotive industries up to small scale machining centres. Although implementing a robotic system can be expensive, it is efficient, productive and accurate; also, the return on investment is assured in a shorter period. Besides the cost of the robot, the main secondary costs of robotizing process are the expenses on peripheral devices like grippers or the end effectors, feeders, tooling, and fixtures [2]. According to Robotiq blog [3], depending on the material, a robot gripper can cost from 10 to several thousand dollars.

The grippers are essential part of a robotic system to manipulate products. In industrial application, most of the tasks are of pick-and-place type where robot picks up a part, manipulates it and puts it in a defined place. According to the Robotics Division Manager of Staubli Corporation in North America [4], 60% of the robotic applications in that region are covered by some sort of handling processes. Therefore, in these types of ap-

* Corresponding author, salman.azim@tuni.fi

plications, grippers are the most important components of the system as they ensure proper gripping of a product and transport it securely without any damages.

Selecting gripper for a certain application can be challenging task. It depends on many factors. The success of choosing suitable gripper depends on its versatility and gripping range.

Nowadays, gripper manufacturers are innovative and produce grippers with multiple gripping solutions, i.e. outside gripping, inner gripping; and gripping objects of both rectangular and cylindrical shapes with a single gripper. Nevertheless, there is always need for universal grippers being able to perform specific tasks. In this article, an approach of selecting proper gripper and the methodology for achieving universal gripping solution are discussed. Furthermore, a use case is analysed by implementing the methodology presented in this article.

2. STATE-OF-THE-ART

In the context of the present article, universal gripping solution does not necessarily represent the idea of using only one gripper to produce company's whole product range, although it would be the primary option. This article focuses on the applications, which concern small batch productions where changes are inevitable. Therefore, flexibility and productivity are the utmost desired qualities of the gripping department. Definition of universal gripping regarding this article is based on readily available gripping system used in the robotic work cell for producing the company's whole product range.

At present, there are already some universal gripping solutions available on the market, based on this definition. For example, automatic gripper changing systems, multiple gripper frames, end of arm tooling (EOAT) etc., are the most commonly used solutions. However, different solutions can have different costs. For example, implementing single gripper with multiple finger options is cheaper than multiple gripper solution. Some of these solutions are readily available and manufactured by many different companies; some solutions can be implemented by in-house engineering for cost minimization. This section of the paper presents state-of-the-art solutions for universal gripping.

2.1. Gripper changing solution

Multiple grippers can be used for gripping different parts. The solution is applicable when parts with different gripping requirements are involved, e.g. pneumatic gripping, magnetic gripping etc. Different grippers are placed in a rack and specific type of gripper is selected according to the specific group of parts being in operation. Changing the gripper will essentially increase the cycle time of the process, as it is unproductive. It also depends on the changing mechanism of the gripper. Two following changing mechanisms are applicable for this type of application:

(1) Automatic tool changer. Automatic tool changers are mounted on the robot wrist on which the gripper is mounted. These are usually disk-type automatic tool changers. Fig. 1 [5] illustrates some common tool changers. The changing disk of the robot is fixed and gripper's work is organized by electrical, pneumatic or electromagnetic control. In case of such modular system, fixing the gripper on the robot takes little time. The cost of such solution is higher as the automatic tool changer increases the cost;

(2) Manual tool changer. In case of manual tool changers, whenever the change of a gripper is needed, it has to be done by human hand. This potentially poses some challenges. A human operator needs to be present while the tool is changed. The cycle time increases as the production is stopped while the changing process takes place. However, using this solution minimizes the cost. Figure 2 [6] is an example of manual tool changer.

Fig. 1. Automatic tool changer [5].

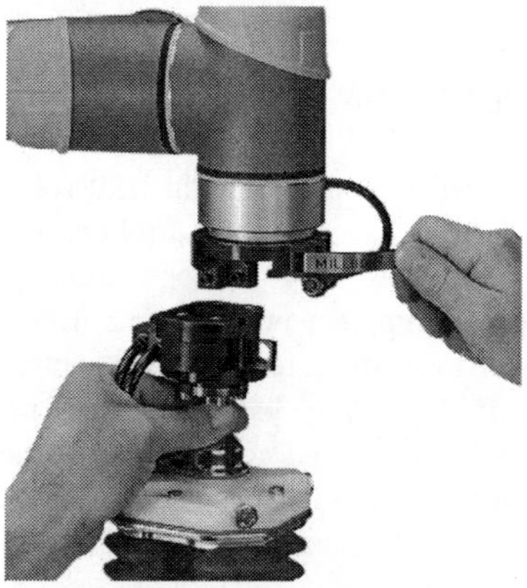

Fig. 2. Manual tool changer [6].

2.2. Multiple gripper frame

In this method, a frame is attached to the wrist of the robot, which can connect two or more grippers at the same time. The grippers remain connected with the robot all the time. Usually, most of the solutions provide attaching two grippers at a time. For more grippers, custom made frames are needed. There are some challenges concerning this type of solutions. First, the load increases as the robot carries the grippers all the time. Second, it decreases the speed of operation. Third, it increases the cycle time. However, the cycle time will be shorter compared to the gripper changing solution. Figure 3 [7] exemplifies the idea of a multiple gripper frame.

2.3. Finger changing solution

Robotic grippers usually have jaws or fingers that interact with components. In case of finger changing solution, multiple finger sets can be used for gripping different parts. This type of solution can be applied only if gripper meets all gripping requirements needed for different kind of parts. Specifications that need attention are: stroke of a gripper, gripping speed; and gripping force.

In case of this method, all necessary finger sets are placed in a rack. Fingers can be changed automatically by automatic screwing and/or detaching, while the robot holds its position. Fingers can also be changed manually, which would increase the cycle time.

This method is used when finger-based grasping is the only applicable choice. This solution is less expensive than the two previous solutions as only one gripper is used for whole spectrum of components.

2.4. Finger dimensioning

This is the most cost effective option of all methods where one gripper can meet the gripping requirements of all parts. In case of this method, only one finger set is used and patterned according to the gripping surface area. This limits gripping of multiple shaped objects. Usually, grippers are sold with some general fingers or without any fingers. Some gripper manufacturers modify the design of the fingers for some additional cost. Modification can also be done in-house either by machining or 3D printing, depending on the material used for the fingers. Machining is possible for steel fingers. Designed fingers can be printed easily due to the development of 3D printing technology. However, 3D designing and printing or milling require time. Determination of gripping area of different parts is also challenging [2]. This method can provide the shortest cycle time.

2.5. End of arm tooling (EOAT)

End of arm tooling is another method of using multiple gripping units at the same time. In case of this method, the robot wrist is connected to a frame, which carries all gripping tools necessary for different steps of the process. This type of system is applied in case of heavy duty robots when the need for different tools is greater than any other system can provide; or gripper needs to carry heavy loads. In practice, these types of solutions are challenging and expensive but they are beneficial due to different gripping tools with multiple control mechanisms connected to the robot at the same time. An example of EOAT is shown in Fig. 4 [8].

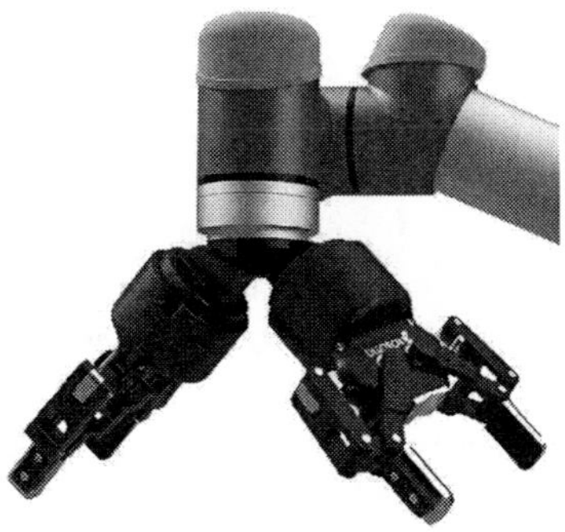

Fig. 3. Multiple gripper frame [7].

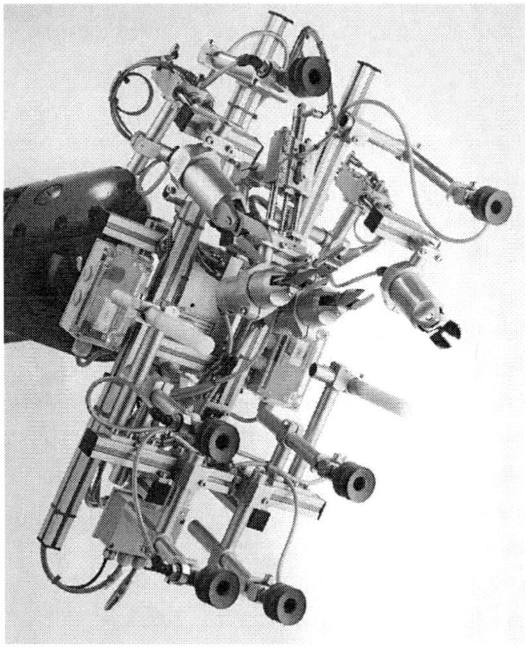

Fig. 4. End of arm tooling [8].

3. GRIPPER SELECTION

At the initial stage of any project, it is suggested to choose suitable gripper according to the task. Gripper selection can be daunting due to huge variety of different options and enormous list of specification sheets. Pham and Yeo [2] describe a part family based and a knowledge based gripper selection processes. From both processes, the first one is more cost-effective while the latter one solves wide range of tasks. In their article, they also present five governing factors for gripper selection: components, tasks, environment of the system, robot, and gripper itself. Each of these factors are listed with some measurable features, which should be taken into consideration while choosing the gripper. These factors are presented below in Fig. 5 [2].

As all these factors depend on the application, the focus of this paper will be on the gripper selection. The part family grouping [2] is followed by additional conclusions. Following steps are proposed for gripper selection, considering the product range of any specific applications:

Step 1 – component study. The aim of the first step is to study the components in detail. The final goal of this step is analysing the gripping surfaces of the components. 3D CAD model of the components can help if physical components are not available. Part family distribution can be made after engineering potential gripping surfaces. This distribution can be done by determining the most common surface areas of parts and dividing them into smallest number of groups. The number of groups will determine the number of grippers required for the application. Other factors which are determined during this step are: component material of gripper fingers, weight of the parts, temperature tolerance, applicable force for the parts; and the orientation of the parts while being in operation. This step will validate the decisions of next steps.

Step 2 – gripper attribute selection. From the results of step one, the important specifications of the gripper can be decided. From numerous specifications of a gripper, the crucial ones are presented below:

- Gripping mechanism. From the results of Step 1, gripping mechanisms have to be determined for all parts. Gripping mechanisms include the types of controllers and the types of tools needed for gripping, e.g. parallel gripping, angular gripping, inner gripping, vacuum gripping, magnetic gripping etc. In this step, control mechanism for the gripper should be determined, as additional equipment like compressed air supply for pneumatic, fluid supply for hydraulic and power supply for electric grippers are essential;
- Stroke width. The gripping surface area of the whole range of parts should be determined in Step 1. Depending on the gripping surface width, from minimum to maximum, the stroke width of the gripper can be selected;
- Gripping force. Gripping force of the gripper depends on several factors. This property is important, as it will guarantee safe handling and transportation of parts. An equation is presented in [9–11] to calculate force for any type of components. The equation is presented below:

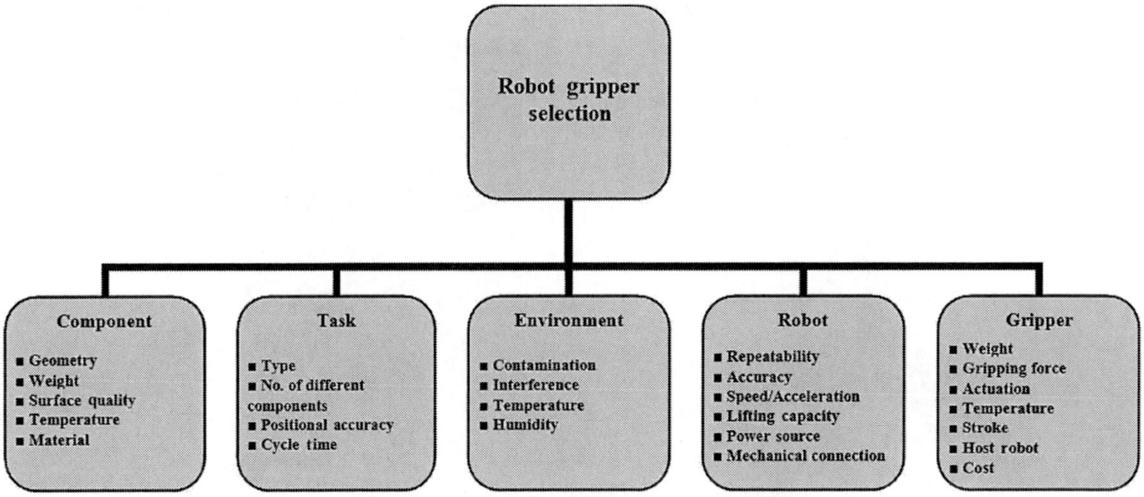

Fig. 5. Governing factors for gripper selection (adapted from [2]).

$$F > m (g + a) / \mu * \text{safety factor}, \quad (1)$$

where, F is the force (N); m is the mass of the product (kg), g is the gravitational acceleration (9.81 ms^{-2}), a is acceleration (ms^{-2}), and μ is the coefficient of friction between gripper finger and the work piece. Safety factor is added according to the application. The total required force will be the sum of the fingers of the gripper. While [9–11] suggests to apply 10 to 20 times more gripping force than the part weight, [11] suggests to add four times more force for friction grip than encompassing grip. They also suggest to consider the finger length and the overhang as it affects the gripping force. On the website [12], there is a list provided for coefficient of friction between different types of materials;

- Gripping speed. Gripping speed is also important as it varies between grippers. Gripping speed can affect the cycle time of the process;
- Repeatability. Repeatability of a gripper is another important property for critical applications. This property ensures the gripper's position accuracy of repetitive task over a certain cycle period;
- Operating temperature. This aspect is important for working environments with high temperature, as grippers can operate in certain temperature range;
- Mounting type. Mounting type of the gripper should integrate with the robot. If mounting type does not match the robot wrist configurations, there will be additional costs.

Step 3 – universal gripping method selection. Following Steps 1 and 2, makes selection of specific kind of universal gripping mechanism from the methods presented in Chapter 2 (State of the art) easier. It is suggested to make measurements and decisions carefully in Steps 1 and 2, as it can help to save costs.

The procedure can be followed using the process flow shown below in Fig. 6.

4. USE CASE ANALYSIS

This chapter analyses a use case for universal gripping method selection. A range of switch mode power supply (SMPS) transformer coils are used for selecting the gripper and achieving universal gripping. For this use case, 54 individual SMPS transformer coils with different sizes, shapes and weights were used. The aforementioned steps were followed to achieve the result:

Step 1. In this step the whole range of parts were studied in detail. The weight, width of gripping surface, material of a part; and orientation of gripping were determined for each part. The weight of the parts varies between 0.02 kg to 0.1 kg. All parts are made of plastic. The gripping surface area varies between 7 mm and 66 mm. Next, all parts were divided into two groups based on the gripping surface. Figure 7 presents two groups.

There are two possible gripping areas for the cylindrical-shaped coils on the left picture: (1) inside gripping by opening the jaws; (2) outside gripping by closing the jaws, objects are grasped from the top.

The rectangular-shaped coils on the right picture can be gripped by their outer frame.

Step 2. As all required parameters of the parts are achieved, the gripper specifications can be determined. The gripping mechanism for all parts can be either pneumatic or electric 2-finger parallel gripper. The first option would be the most cost-effective solution. If the first group has to be gripped from inside, 3-finger gripper is required. The stroke width needed for the gripper is between 4–69 mm as the minimum and maximum width of gripping surface varies between 7–66 mm. Gripping force of all parts varies between 16.33 N and 32.66 N, considering 20 times more force than the work piece weight. The coefficient of friction is considered 0.3 for plastic to plastic. Therefore, based on given data, Robotiq 2F-85 gripper with adjustable fingers was selected, with a stroke of 85 mm, and force range from 20 to 235 N. It is also compatible with the previous robot UR5. This gripper is controlled by electrical power source and moves fingers by servomotor;

Step 3. To achieve universal gripping, finger-dimensioning method was used. This would be the most inexpensive solution with shortest cycle time. The ranges of parts also fall in 2 groups, which can be achieved by designing fingers that can grasp both circular outer

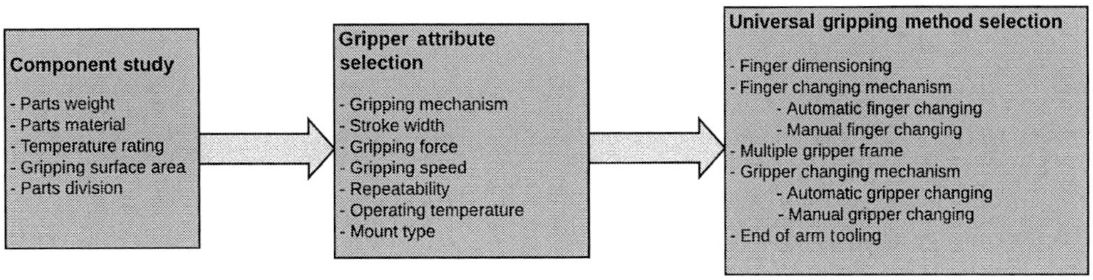

Fig. 6. Process flow for achieving universal gripping.

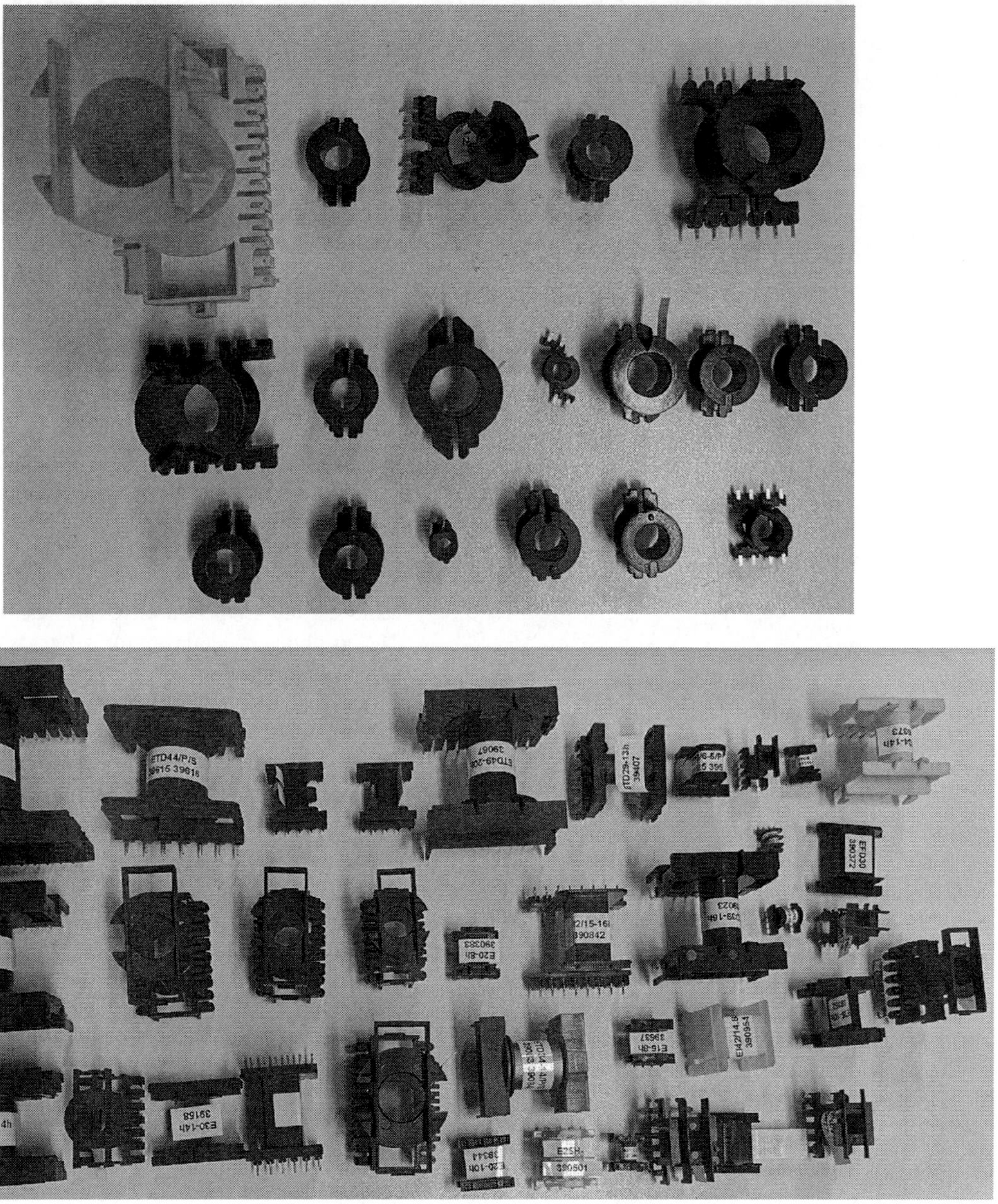

Fig. 7. Dividing parts according to the gripping surface area.

frame of the first group and rectangular outer frame of the second group.

SolidWorks software was used for 3D designing and Prusa printing machine for 3D printing. The outcome is presented in Fig. 8.

This finger configuration was achieved through some trial and error; and could achieve the gripping of both part groups. Some pictures of the gripped parts from both families are provided in Fig. 9.

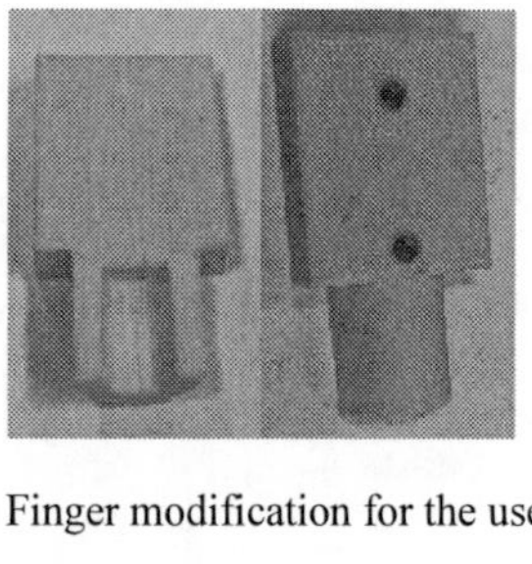

Fig. 8. Finger modification for the use case.

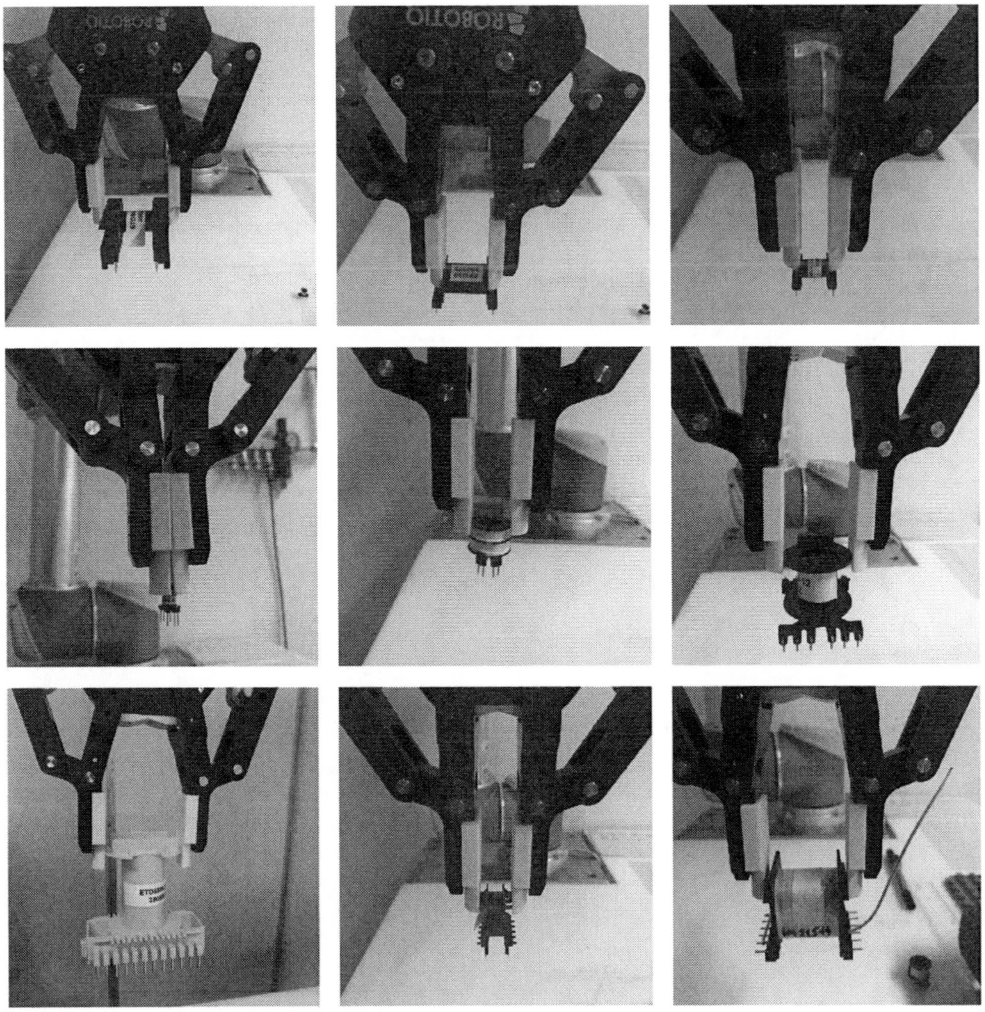

Fig. 9. Universal gripping for the use case using finger modification method.

5. CONCLUSIONS

Competition on the global market motivates companies to work out cost effective solutions with increased efficiency. As the prices of industrial robots are falling, more and more applications are robotized. This was not possible few years ago. Applications of small batches of products with diverse range need inexpensive solutions for full-scale automation. Following the discussed methodology can make gripper selection easier for applications that involve parts with different shapes, sizes and materials. Focusing on different steps and making proper decisions can save costs and achieve universal gripping solution, thus minimizing the overall costs of robotizing any processes. At present, there isn't any specific gripper selection methodology and relevant information for gripper selection requires expertise. This article can help to select proper gripper and to achieve cost-effective universal gripping system in the inicial phase of establishing a robot-based production.

ACKNOWLEDGEMENTS

The publication costs of this article were covered by Tallinn University of Technology and the Estonian Academy of Sciences.

REFERENCES

1. Appleton, E. and Williams, D. J. *Industrial robot applications*. Springer Science & Business Media, 2012.
2. Pham, D. T. and Yeo, S. H. Strategies for gripper design and selection in robotic assembly. *Int. J. Prod. Res.*, 1991, **29**(2), 303–316.
3. Bouchard, S. Robot gripper: How much does it cost? August 17, 2014. https://blog.robotiq.com/bid/58812/Robot-gripper-How-much-does-it-cost (accessed 2018-12-17).
4. Camillo, J. Options for Pick and Place. *ASSEMBLY*, November 5, 2018. https://www.assemblymag.com/articles/94544-options-for-pick-and-place?v=preview (accessed 2018-12-17).
5. Changing: Quick Change with System. *Schunk.com*. https://schunk.com/de_en/gripping-systems/category/gripping-systems/robot-accessories/changing/ (accessed 2018-12-17).
6. Millibar manual tool changer. https://www.universal-robots.com/plus/accessories/millibar-manual-tool-changer/ (accessed 2018-12-17).
7. Photos & Videos. *Robotic.com*. https://robotiq.com/resource-center/media (accessed 2018-12-17).
8. Tooling. https://eoat.net/tooling/ (accessed 2018-12-17).
9. Selection Guide (Gripping Force) Technical Reference/Information. http://www.intelligentactuator.com/partsearch/robocylinder/appndx74_Model_Selection_by_RCP2_Gripper.pdf (accessed 2018-12-17).
10. Bélanger-Barrette, M. Robotic End effectors – Payload vs Grip Force. January 29, 2014. https://blog.robotiq.com/bid/69524/Robotic-End effectors-Payload-vs-Grip-Force (accessed 2018-12-17).
11. Robotic Gripper Sizing: The Science, Technology and Lore. http://www.grippers.com/size.htm (accessed 2018-12-17).
12. Coefficient of friction, Rolling resistance and Aerodynamics. http://www.tribology-abc.com/abc/cof.htm (accessed 2018-12-17).

Universaalse haaramislahenduse rakendamise metoodika robotrakendusele

Mohammed Salman Azim, Andrei Lobov ja Artem Pastukhov

Viimastel aastatel on edusammud robootika valdkonnas olnud kasulikud eelkõige töötleva tööstuse jaoks. Korduvaid, monotoonseid ja mittekeskkonnasõbralikke tegevusi pannakse üha rohkem tegema robotid, mis võivad töötada paralleelselt inimestega, tehes nende töö kergemaks. Konkurentsivõimeline turg motiveerib robotitootjaid tagama investeeringute tasuvuse väga lühikese aja jooksul. Oluline arendusvaldkond on robotite haaratsid ja peamised probleemid selles valdkonnas on seotud erinevate toodete haaramisega ühe haaratsiga. Käesolevas artiklis on käsitletud üht lähenemisviisi ja metoodikat toote variatsioonide käsitsemiseks ühe haaratsiga.

Proceedings of the Estonian Academy of Sciences,
2019, **68**, 4, 421–428
https://doi.org/10.3176/proc.2019.4.12
Available online at www.eap.ee/proceedings

MANUFACTURING
ENGINEERING

Granular jamming based robotic gripper for heavy objects

Jesse Miettinen*, Patrick Frilund, Iiro Vuorinen, Petri Kuosmanen, and Panu Kiviluoma

Department of Mechanical Engineering, Aalto University, Sähkömiehentie 4, 02150 Espoo, Finland

Received 19 November 2019, accepted 26 November 2019, available online 29 November 2019

Abstract. Moving heavy objects with overhead cranes requires the operator to fasten the object to a hook with ropes or chains. This is a time-consuming process, which could be avoided by using universal grippers that can lift objects of any shape. This study was conducted to find if a universal gripper, based on granular jamming, can be used for crane scale applications. Maximum lifting capacity of granular jamming grippers was analytically evaluated and experimentally tested with various material combinations. Objects with different shapes, sizes and weights were successfully lifted with selected gripper configurations. The results showed that grain size and grain compressibility both affect the performance of the gripper. It was demonstrated that in order to efficiently lift heavy objects with granular jamming, the granular material has to be compressed sufficiently. Pressure difference between environment and the sealed pouch, filled with granular material, has to be correct. With this setup, gripper based on granular jamming was able to lift objects with various shapes; and weights up to 120 kg.

Key words: granular jamming, crane, lifting, universal gripping.

1. INTRODUCTION

The technology for lifting heavy objects has always relied on strong hooks and some manual labour for attaching an object to the hook in various ways. Usually, in case of lifting objects with the overhead crane, the objects are attached to the hook with chains or ropes. This system, having two mass nodes attached to ropes, is slow and prone to double pendulum dynamics [1]. Also, this system requires workers to work constantly in dangerous environment around the hook. Usually a person has to attach objects manually. Removing the attachment step could save a lot of time and increase safety of the work environment.

Gripping technology in robotics has recently taken big steps towards universal gripping. Universal grippers can lift objects with variable shapes [2]. Current gripping technologies can be divided into three different categories, enabling grasping by: (1) actuation; (2) controlled stiffness; and (3) controlled adhesion. Actuation covers all traditional gripping technologies, such as different rigid jaw grippers and stiff grippers driven by complex algorithms. It also covers the compliant materials deformed during gripping process [3]. The problem with gripping by actuation is that it can be rather clumsy with various shapes and can even cause harm with applied forces [4].

Adhesion-based grippers are able to grip objects with a shear force that is applied on the lifted object. This kind of lifting method can be used to lift large, deformable and easily breakable objects. One way of making an effective adhesive gripper is to put very thin and adhesive micro-sized planes or hairs on the gripping surface of the gripper. When the gripping surface has been placed on the object and applied with shear force, the contact area between the object and the bending micro planes grows. For example, in the nature gecko uses this method when walking on vertical walls. This technique can be used to lift objects in the range of few kilograms, without squeezing them [4].

The idea behind gripping by controlled stiffness is setting the gripper's structure in its soft configuration for enveloping the target object to be grasped [3]. The

* Corresponding author, jesse.miettinen@aalto.fi

gripper is then stiffened so that it grips the target object. One example of this is granular jamming where a sealed and airtight pouch is filled with granular material. When the pouch is at its soft configuration, it can be deformed easily and its contents has fluid-like behaviour in macro-scale. After sucking out the air from the pouch, the granules get jammed and the pouch stiffens to its deformed shape around the target object [5]. The granular jamming process can even be sped up by adding positive pressure during deformation and releasing stages of the gripping process [6]. Stiffness can also be controlled by alloys that change their phase from solid to liquid, depending on the temperature. There are also some fluids that respond to electric and magnetic fields by changing their viscosity [3].

Developments in industrial robotics have gained promising results using granular jamming for lifting objects [3]. For example, Empire Robotics has successfully lifted objects with the mass of around 10 kg with a 16.5-cm-diameter-gripper [7]. Thus, this technology might be viable option for substituting some of the crane lifting that use ropes; and for lifting objects weighing less than few hundreds of kilograms. However, the maximum capacity for such kind of lifting method is still unknown. This study documents the scalability of the gripper based on granular jamming by scaling the size of the pouch and testing different granular materials.

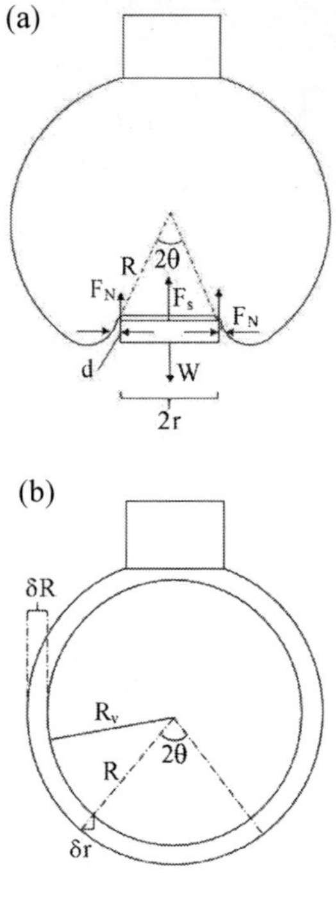

Fig. 1. (a) Free body diagram of the gripper lifting a cylinder; (b) Volume change of the gripper moves each point of the pouch in horizontal direction, change of the radius is **δr**. The large circle represents the cross-sectional side view of the gripper in its initial configuration and the smaller circle represents the gripper when the air has been evacuated.

2. MODEL FOR LIFTING CYLINDRICAL SHAPES

The gripper consists of a large pouch filled with granular material, a body to support the pouch, and a vacuum pump. The gripper assembly is lifted on top of a cylindrical object with radius r and height d (Fig. 1a). The lifting capacity of a gripper based on the granular jamming is hard to estimate as it depends on the shape of the object. Below, the maximum lifting capacity for lifting a cylindrical-shaped object is estimated analytically.

When air has been evacuated from the pouch, the change of the radius is δR. This results in decrease of the radius of the pouch's horizontal cross-section, δr (Fig. 1b). It is assumed that the radius of a cylinder does not change, i.e. the lifted object has a stiffness much larger than that of the granular material. The normal force that hinders contraction can be calculated from the following relationship:

$$\delta r = 0 \Rightarrow \varepsilon_{rr} = 0, \tag{1}$$

and using Hooke's law we get

$$F_N = \frac{\delta r}{r} A_0 E. \tag{2}$$

Slipping occurs when the pressure inside the low-pressure pocket between the top face of the cylinder and the membrane is higher than gripping pressure at the walls of the cylinder. The following relationship holds

$$P_v = \frac{F_N}{A_0}. \tag{3}$$

The suction force caused by the low-pressure pocket is

$$F_s = P_v A_c, \tag{4}$$

where A_c is the area of the top of the cylinder. Now putting these together gives

$$F_s = F_N \frac{A_c}{A_0} = F_N \frac{r}{2d}. \tag{5}$$

The change in the cross-sectional radius is

$$\delta r = \delta R \sin(\theta). \tag{6}$$

Since the angle θ depends on the height of the cylinder inside the gripper, the total force can be found by summing all the differential force elements along the height of the cylinder. The maximum lifting weight of the gripper is then

$$W_{max} = F_s + F_N,$$

$$W_{max} = 2\pi\mu E\delta R\int_0^d \frac{r}{\sqrt{(R-h)^2 + r^2}}dh\left(1+\frac{r}{2d}\right), \quad (7)$$

where E is the Young's modulus of the granular material, R is the radius of the gripper, r is the radius of the cylindrical object, d is the height of the cylindrical object, h is the height of the cylindrical object, μ is the coefficient of static friction, and δR is the change of the gripper's radius.

The change in radius of the gripper can be written in terms of the volumetric strain:

$$W_{max} = 2\pi\mu ER\left(1-\sqrt[3]{1-\frac{\delta V}{V}}\right) \times\left(\tan^{-1}\left(\frac{R-d}{r}\right)\right)\left(\frac{1}{r}+\frac{1}{2d}\right). \quad (8)$$

It can be seen that variables, which can be affected by design of the gripper, are: *radius of the gripper*, *volumetric strain* and the *Young's modulus* of the granular material. The volumetric strain is proportional to the pressure difference between the inside and outside of the gripper membrane (bulk modulus). Therefore, a sufficiently powerful vacuum pump is essential for the gripper to provide high gripping forces. The choice of granular material is also essential for the gripper to provide high gripping forces. Young's modulus of the granular material must be high enough to be able to compress and deform around various objects.

It should also be noted that the gripping strength depends on the geometry of the gripping object, not only on the contact surface area but also on the contact angle. The contact angle was assumed to be 90 degrees in these calculations, i.e. the object's contact surface is perpendicular to the ground.

Figure 2 illustrates the maximum lifting capacity for the cylinder with radius of 15 cm and height of 10 cm as

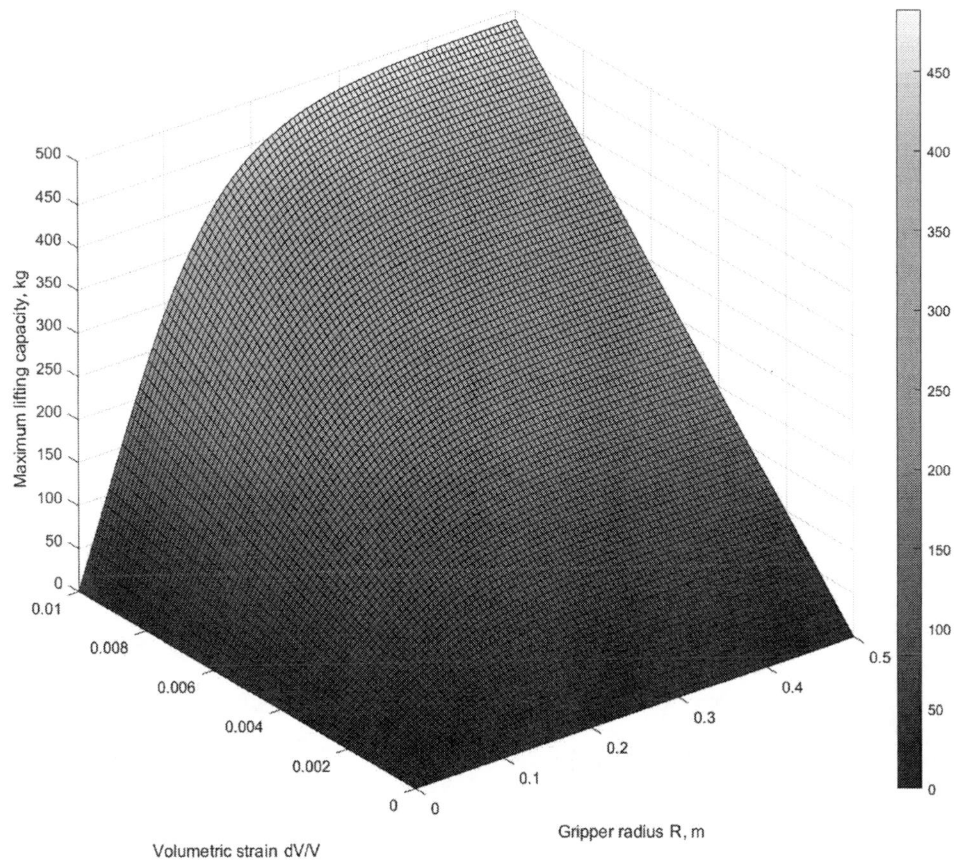

Fig. 2. Design variables vs maximum lifting capacity for a cylinder with r = 0.15 m and d = 0.1 m. The volumetric strain was assumed to be dV/V = 0.004 and the radius of the gripper R = 0.325 m. Value of the volumetric strain was taken from previous research [5]. Plot was produced with MatLab.

a function of volumetric strain and radius of the gripper. Granular material used in the calculation is coffee and the values for the stiffness, friction coefficient and volumetric strain were taken from previous research [5].

3. EXPERIMENTAL DESIGN

In order to gain some knowledge about the effect of granules' size on the gripping force, wooden pellets were chosen as the first test material (Table 1). The same wooden pellets in a ground form were chosen to be the second material, to achieve smaller particle size. The wooden pellets were chosen for their relatively advantageous stiffness-lightness ratio. The third granular material used for the study was plastic granules, which were even lighter and stiffer than wooden pellets. The fourth test material was sand, which has very high stiffness and very high density. Fifth material was rubber granules for their elasticity. The rubber granules were similar to the ones used in artificial football fields. All granules are shown in Fig. 3.

A typical exercise ball was chosen to be the surface material of the gripper (Table 2). The sheer weight of the granular material inside the gripper's pouch can cause great stresses to the surface material and break it. Thus, a custom pouch was made of carbon and glass fibre composite. The composite pouch was then sealed with silicone coating.

The rest of the system includes a custom rigid structure connecting the pouch with the hook of the crane. Also, a suction system was connected to the rigid body in order to create pressure difference between the contents of the membrane and environment. The vacuum machine used was Robinair single stage vacuum pump achieving an ultimate pressure of 0.2 bar. The pump and outlet valve in the frame body were controlled by the smartphone application via Bluetooth with an Arduino UNO, a relay and a servo motor. The final setup can be seen in Fig. 4.

4. TESTING ENVIRONMENT

Testing was done with multiple different combinations of granules and membrane surface materials. The combinations are listed in Table 3. Every combination of the test material was put through the same sequence of lifting tasks.

Lifting tasks were performed by placing the lifted object on the ground and lowering the gripper on top of the object. The gripper was lowered until it was no longer hanging from the crane but rather resting on top of the test object. The absolute pressure inside the gripper was lowered to approximately 0.6 bars using the 0.19 kW vacuum pump.

The first task was to lift a 1 kg plastic bucket (Table 4) to make sure the vacuum pump was working

Table 1. Granular material parameters

Granule material	Granule size, mm	Density, kg/m^3	Stiffness, GPa
Wooden pellets	d = 6+/–0.2; l = 15.6+/–5	696	1
Ground wooden pellets	d = 6+/–0.2; l = 5 +/– 2.5	469	1
Plastic granules	2.5	602	3
Sand	0.5–1.5	1559	50
Rubber granules	2.5	535	0.1

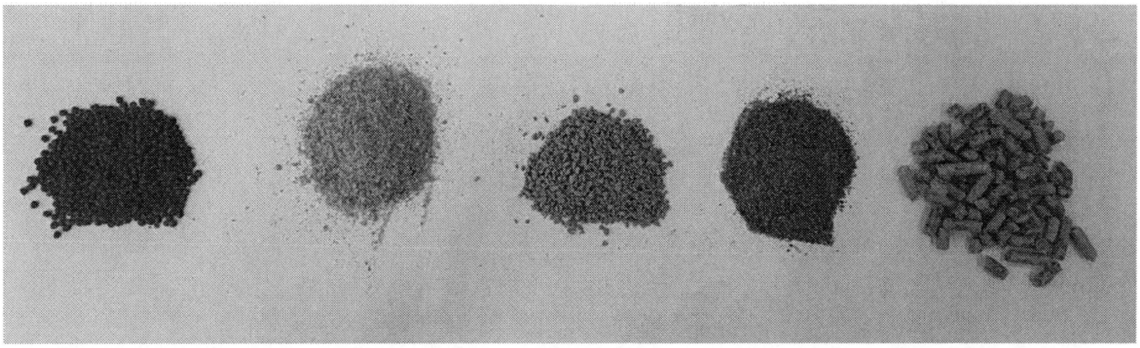

Fig. 3. Five different types of granular materials starting from the left: plastic granules, ground wooden pellets, rubber granules, fine sand; and wooden pellets.

Table 2. Pouch surface material parameters

Reference	Materials	Tensile strength	Elasticity	Thickness, mm	Diameter, cm
Exercise ball 1	PVC	Low	High	1.5	65
Exercise ball 2	PVC	Low	High	1.5	85
Custom ball	Silicon coating, carbon fibre/glass fibre layer	High	Low	2–4	65

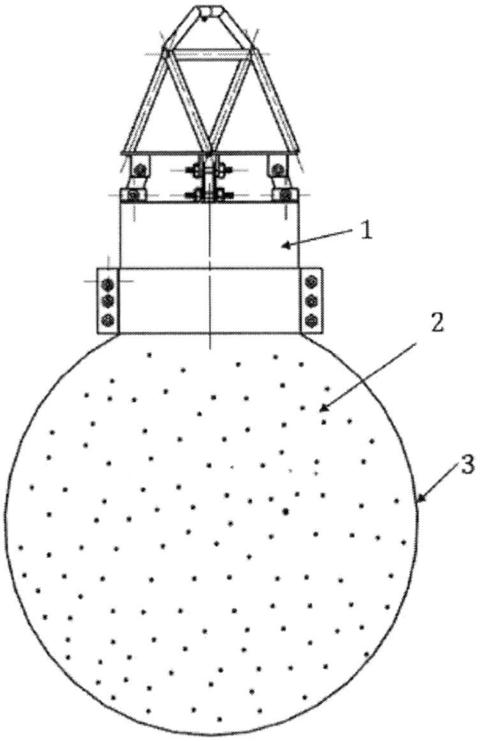

Fig. 4. (a) Gripper configuration consisting of the following components: the frame body (1); granular material (2); and the gripper's surface material (3). (b) Final gripper assembly with vacuum pump mounted on top.

Table 3. Combinations of tested materials' parameters

Combination	Granular material	Pouch surface
1	Wooden pellets	Exercise ball 1
2	Ground wooden pellets	Exercise ball 1
3	Plastic granules	Exercise ball 1
5	Sand	Exercise ball 1
6	Wooden pellets	Exercise ball 2
7	Wooden pellets	Custom ball
8	Rubber granules	Exercise ball 1

properly. The following three tasks included lifting three various objects weighing roughly 5–10 kg, to evaluate the grip on complicated shapes. These three different objects were: a stool, a fire extinguisher and a gym ball. The next task was to lift a trash bin with sand and metal filling, weighing 30 kg. The final object was a 120 kg steel cylinder with a hook attachment protruding from the top. Test objects, except for the 1 kg plastic bucket, are displayed in Figs 5a–5e. Lifting tasks were classified either as a success or a failure.

A task was successful if the gripper could lift the object and hold it up for 30 seconds. Lifting task could be repeated up to ten times before it was classified as a failure. Ten trials were given since finding out the optimal grip requires multiple trials. One successful lift was enough to classify the test successful.

Table 4. Test objects and dimensions

Test object No.	Object	Radius, cm	Height, cm	Weight, kg
1	Plastic bucket, standing	10	30	1
2	Stool, standing	32	57 (2.5 cm contact area)	5
3	Gym ball	28	–	8
4	Fire extinguisher, laying	15	42	9
5	Trash bin filled with sand and metal weight, standing	27	30	30
6	Steel cylinder, standing	25	35	120

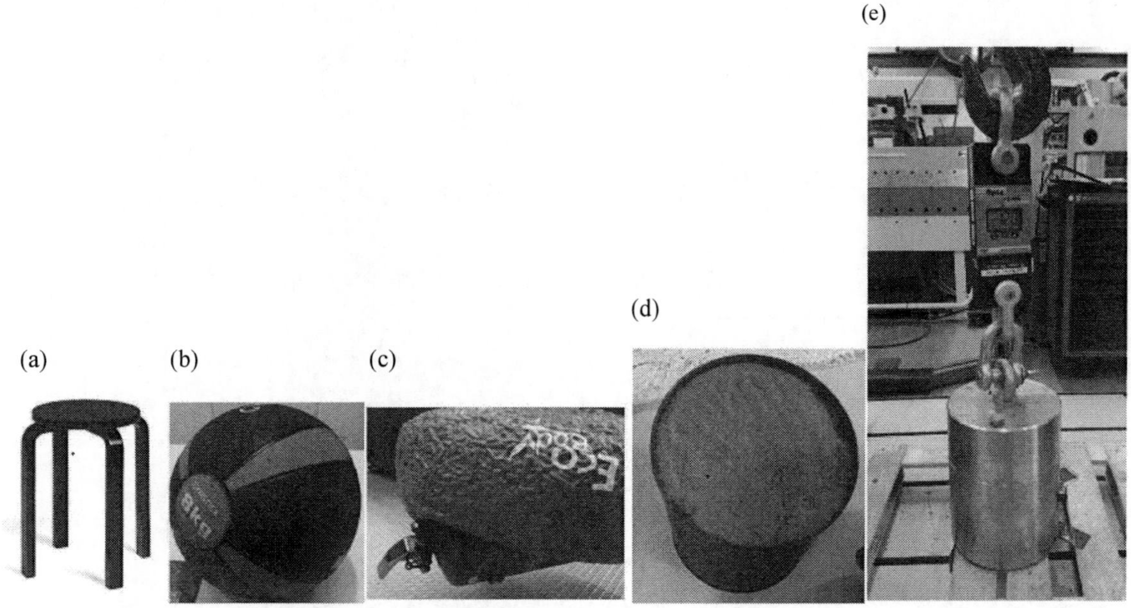

Fig. 5. Test objects: (a) a stool (picture does not present the actual test object); (b) gym ball; (c) fire extinguisher; (d) trash bin filled with sand; and (e) steel cylinder. Plastic bucket is not shown in the figure.

5. RESULTS

Tests of the granular materials and membrane materials were performed according to the previous section. The results of the tests are listed in Table 5.

In general, only 0.2 bar pressure difference was usually enough to generate sufficiently tight gripping for lifting. 120 kg cylinder required more than 0.5 bar pressure difference for successful lifting. The heavier the object, the higher was required pressure difference between environment and the pouch.

The tests made with smaller exercise ball and wooden pellets showed promising results. This combination was able to complete first five tasks. However, tests to lift

Table 5. Test results for each combination of granular material and surface material of the gripper

No. of test object and the combination	1	2	3	4	5	6
1 Exercise ball 1, wooden pellets	Success	Success	Success	Success	Success	–
2 Exercise ball 1, ground wooden pellets	Success	Success	Success	Success	Success	Success
3 Exercise ball 1, plastic granules	Success	Success	Success	Success	Success	Failure
4 Exercise ball 1, sand	Success	Success	Failure	Failure	Failure	Failure
5 Exercise ball 2, wooden pellets	Failure	Success	Failure	Failure	Failure	Failure
6 Custom ball, wooden pellets	Failure	Success	Failure	Failure	Failure	Failure
7 Exercise ball 1, rubber granules	Success	Success	Success	Success	Success	Success

the 120 kg cylinder were not carried through as it took multiple trials to lift successfully objects 3, 4, and 5. The second test was done with the smaller exercise ball and the ground wooden pellets. This combination gave better results than the first one, and completing the task of lifting the first five tasks was successful already with the first trial. The gripper was also able to lift the 120 kg steel cylinder during ten trials.

Plastic granules performed well, lifting successfully the first five tasks within given ten trials. However, this combination was not able to lift the 120 kg cylinder. Sand showed poor results, completing successfully only first two tasks. The larger exercise ball and the custom ball showed poor performance being capable of lifting only the 5 kg stool.

The best combination seemed to be smaller exercise ball filled with rubber granules. All lifting tasks were marked successful with this combination already with the first trial.

Fig. 6. Large wrinkles forming on the gripper's surface, when the gripper with ground wooden pellets as granular material was evacuated of air.

6. DISCUSSION

Granular materials with higher compressibility gave the best results. As mentioned, the maximum lifting capacity depends on the volumetric strain of the gripper. The volumetric strain increases if the volume of the granular material can be compressed. This results in higher gripping forces and thus a larger lifting capacity. If the material is compressible it also deforms more easily around the surface of the lifted object. This in turn results in a larger contact surface area between the object and the gripper and therefore also larger gripping forces.

The most compressible granular materials were the ground wooden pellets and the rubber granules. This was indicated by the large wrinkles formed on top of the gripper's surface after evacuation of the air (Fig. 6). These materials gave also the best test results, supporting analytical model. It should be noted that sand was the poorest granular material, and it was also the least compressible material.

Comparing test results with the analytical solution it has to be kept in mind that granules were assumed to be similar to the ground coffee. The material, which resembles coffee the most is the ground wooden pellets. This material has similar size of granules and it is compressible. Equation (8) gives a maximum lifting capacity of approximately 270 kg for objects that have similar shape with the test object 2. This is above the weight of the heaviest test object.

Considering combination 2, it is most likely that the maximum lifting capacity for objects with similar shape as object 6 is somewhere near 120 kg. Object 6 was successfully lifted only once with this combination. However, combination 7 successfully completed the same task with the first trial. This might be due to higher compressibility of the granular material. The maximum lifting capacity of combination 7 can be relatively near to the analytical approximation of 270 kg. However, finding the maximum lifting capacity for combination 7 requires further experiments.

According to Eq. (8), the gripper's lifting capacity is proportional to the radius of the gripper. Larger radius should give a larger lifting capacity. The tests showed that the larger exercise ball gave poor performance quality, as it was only capable of lifting the stool. It must be noted that only 50% of the larger exercise ball was filled up with granules while performing the tests. Other balls were filled up to around 80%, which is enough to leave enough space for free movement between the granules before jamming. Further testing with gripper radii should be done to verify the relationship in Eq. (8).

Among the same type of granular materials, the one with the smallest granule size performs the best. This can be seen when comparing tests with wooden pellets and ground wooden pellets. It took multiple trials to successfully lift the first five objects with combination 1, and only one trial per object with the combination 2.

Previous studies [5] have shown that it is possible to use granular jamming to lift objects with various shapes. Current study showed that it is possible to use granular jamming in crane-scale applications. This kind of universal gripper based on the granular jamming would be suitable for lifting objects with various shapes and sizes, weighing up to 120 kg. Such a gripper could be used in machine shops or warehouses where heavy objects with irregular shapes are moved frequently.

ACKNOWLEDGEMENTS

The publication costs of this article were covered by Tallinn University of Technology and the Estonian Academy of Sciences.

REFERENCES

1. Kim, D. and Singhose, W. Performance studies of human operators driving double-pendulum bridge cranes. *Control Eng. Pract.*, 2010, **18**(6), 567–576. https://doi.org/10.1016/j.conengprac.2010.01.011
2. Felip, J. and Morales, A. Robust sensor-based grasp primitive for a three-finger robot hand. In *Proceedings of the 2009 IEEE/RSJ International Conference on Intelligent Robots and Systems, St. Louis, MO, USA, October 10–15, 2009*. IEEE, 2009, 1811–1816. https://ieeexplore.ieee.org/document/5354760
3. Shintake, J., Cacucciolo, V., Floreano, D., and Shea, H. Soft Robotic Grippers. *Adv. Mater.*, 2018, **34**(29). https://doi.org/10.1002/adma.201707035
4. Hawkes, E. W., Jiang, H., Christensen, D. L., Han, A. K., and Cutkosky, M. R. Grasping Without Squeezing: Design and Modeling of Shear-Activated Grippers. *IEEE Trans. Rob.*, 2018, **34**(2), 303–316.
5. Brown, E., Rodenberg, N., Amend, J., Mozeika, A., Steltz, E., Zakin, M. R., Lipson, H., and Jaeger, H. M. Universal robotic gripper based on the jamming of granular material. *PNAS*, 2010, **107**(44), 18809–18814. https://doi.org/10.1073/pnas.1003250107
6. Amend, J. R., Jr., Brown, E., Rodenberg, N., Jaeger, H. M., and Lipson, H. A Positive Pressure Universal Gripper Based on the Jamming of Granular Material. *IEEE Trans. Rob.*, 2012, **28**(2), 341–350. https://doi.org/10.1109/TRO.2011.2171093
7. Amend, J., Cheng, N., Fakhouri, S., and Culley, B. Soft Robotics Commercialization: Jamming Grippers from Reasearch to Product. Soft Rob., **3**(4), 213–222. https://doi.org/10.1089/soro.2016.0021

Teratihil põhinev haarats raskete esemete teisaldamiseks

Jesse Miettinen, Patrick Frilund, Iiro Vuorinen, Petri Kuosmanen ja Panu Kiviluoma

Kraanadega raskete esemete teisaldamiseks peab operaator kinnitama teisaldatava eseme köite või kettide abil kraanakonksu külge. See on aeganõudev protsess, mida saaks vältida universaalsete haaratsite abil, mis võimaldavad tõsta mistahes kujuga esemeid. Uuring viidi läbi selleks, et teada saada, kas graanulite teratihil põhinevat universaalhaaratsit saab kraanade jaoks kohandada. Graanulite teratihil põhineva haaratsi maksimaalset tõstevõimet hinnati analüütiliselt ja katsetati erinevate materjalide kombinatsioonidega. Erineva kuju, suuruse ja raskusega esemed tõsteti valitud haaratsikonfiguratsioonide abil edukalt üles. Tulemused näitasid, et nii graanuli suurus kui ka selle kokkusurutavus mõjutavad haaratsi jõudlust. Tõestati, et teratihtimise abil raskete esemete tõhusaks tõstmiseks peab graanulimaterjal keskkonna ja graanulimaterjali kinnise tasku tekitatud rõhkude erinevusest tingitult olema piisavalt kokku surutud. Katseseadistuses suutis teratihil põhinev haarats tõsta erineva kuju ja massiga kuni 120 kg raskusi esemeid.

AUTHOR INDEX

Al-Fandi, L. 376
Alfailakawi, B. 376
Alsubaiei, H. 376
Azim, M. 413
Bondarenko, Y. 364
Eerme, M. 371
Frilund, P. 421
Igavens, M. 356
Kaganski, S. 371
Kangru, T. 383
Karaulova, T. 348, 356
Karjust, K. 401
Khudhair, S. 376
Kiolein, I. 407
Kiviluoma, P. 421
Kruuser, K. 401, 407
Kuosmanen, P. 421
Kuts, V. 364
Lobov, A. 413
Mahmood, K. 348, 356, 383
Malayjerdi, E. 395
Miettinen, J. 421
Modoni, G. 364
Obaid, A. 376
Otto, T. 348, 364, 383, 389
Pastukhov, A. 413
Pataraia, T. 395
Rassolkin, A. 389
Riives, J. 383, 407
Sacco, M. 364
Sell, R. 389, 395
Shevtshenko, E. 348, 356
Strods, G. 356
Tahemaa, T. 364
Tandzegolskiene, I. 356
Tsukrejev, P. 401, 407
Tungel, E. 371
Tutlys, V. 356
Valjaots, E. 395
Vuorinen, I. 421
Wang, R. 364, 389